Mathematics Education for a Changing World

Stephen S. Willoughby

Association for Supervision and Curriculum Development
Alexandria, Virginia

The Author: Stephen S. Willoughby has taught all levels from 1st grade through graduate school. He is Professor of Mathematics at the University of Arizona and has previously taught at New York University and the University of Wisconsin. Willoughby is Past President of the National Council of Teachers of Mathematics and former Chairman of the Council of Scientific Society Presidents. He is senior author of *Real Math*, published by Open Court; advisor for SQUARE ONE TV; and director of a National Science Foundation project using calculators and computers in grades 1–6. He is active in many other professional groups.

Printed in the United States of America. Typeset on Xerox™ Ventura Publisher 2.0.

Ronald S. Brandt, *Executive Editor*
Nancy Modrak, *Managing Editor, Books*
Carolyn R. Pool, *Associate Editor*
Ginger R. Miller, *Associate Editor*
Lars Kongshem, *Editorial Assistant*
Stephanie Kenworthy, *Assistant Manager, Production Services*
Valerie Sprague, *Desktop Typesetter*
Simeon Montesa, *Graphic Designer*

$11.95
ASCD Stock No. 611-90100
ISBN: 0-87120-175-5

Library of Congress Cataloging-in-Publication Data.

Willoughby, Stephen S.
    Mathematics education for a changing world/Stephen S Willoughby
        p.   cm.
    Includes bibliographical references.
    ISBN 0-87120-175-5
    1. Mathematics—Study and teaching.   I Title
    QA11.W563   1990
    510'.71—dc20

                                                           90-49407
                                                              CIP

Mathematics Education for a Changing World

Foreword

My earliest memories of mathematics are happy ones. Mathematics—like reading—was powerful. Understanding mathematics separated you from the "babies" in our family. When the secret of the "nines" multiplication facts was passed on to you, you became one of an elite group, and it sure was fun when it was your turn to recite your timestables through the "ninezers" without a mistake.

Other kids may have groaned about story problems for homework, but I inwardly cheered because good story problems were like magical puzzles to be unlocked with a little reason and a little figuring. From the beginning, the expectation was that my siblings and I would all understand the mathematics world and be facile with figures. We often watched my dad use estimation to determine the amount and sizes of lumber needed to build a garage or remodel an existing building. My siblings and I were admonished as we matured for needing paper "to figure," except when doing schoolwork. On schoolwork it was imperative that we showed both our answers and our methods—no matter how simplistic and redundant.

Today's mathematics, as described in this book, would be quite in tune with the reality of my home-taught mathematics. The standards issued by the NCTM/NCSM groups carry with them the obligation to make mathematics come alive for youngsters, to make it useful in real-world problem solving and communication. The standards emphasize the connectedness of each of the branches of mathematics and underline its relationship to science and technology. The standards also propose teaching estimating, graphing, statistics, and probability at earlier stages than is now common.

This text turns these new standards into a practical reality for teachers and school administrators. Seldom is book learning so stimulating as it is here, thanks to the standards and Willoughby's fresh approach.

How long must we wait for this commonsense approach to become a reality in the classroom?

Donna Jean Carter
ASCD President, 1990–91

1

Why Change?

Teachers who have watched self-proclaimed experts lead us through the "new math" programs of the 1950s and 1960s, the "back-to-basics" movement of the 1970s, and the "critical thinking" movement of the 1980s might be excused—indeed, might be commended—for looking with a somewhat jaundiced eye at proposals for still more changes. Are such changes justified?

Our world is becoming more mathematical. We are constantly surrounded by mathematical situations and are regularly required to make mathematical decisions. These decisions require number sense, estimations skills, ability to analyze data intelligently, knowledge of two- and three-dimensional geometry, knowledge of probability, and many other abilities not often taught in school.

Most people who find themselves in a situation that requires mathematics either don't recognize that good decisions depend on mathematical thought or don't make the best decisions because they are unable or unwilling to think mathematically. The fast-food clerk who returns $98.16 change to a customer who puts down only a ten-dollar bill, the young mother who believes she's certain to have a girl next time because she's had three boys already ("these things even out in the long run"), and the person who believes that a 40 percent chance of rain on Saturday and a 60 percent chance of rain on Sunday guarantee a 100 percent chance of rain on the weekend are all examples of people failing to think mathematically when they should. Most such examples are more subtle.

The public's first intimation that there might be something seriously wrong with mathematics education in the United States came with publication of the results from the first International Study in Mathematics Education in 1964. Japan appeared to have done much better than other countries. The United States appeared to have done worse.

Editorial writers, politicians, and others viewed these results with alarm for a few months, but were soon pacified by knowledge that intercountry comparisons were essentially impossible to draw from the study because of the nature of the tests and the populations studied. If the tests were skewed more toward the curriculum in country A than country B, country A could be expected to do better on the tests. Because of a strong commitment to universal education in some countries, including the United States, the top 90 percent of students in those countries were compared with the top 20 percent in other countries.

Insofar as concern over the apparent weakness in mathematics education in the United States had any effect, it was used to condemn and help kill the "new mathematics" that had just begun to influence the curriculum. Because virtually all the children tested in the 1964 study had been taught mostly, or only, "old" mathematics, there seems to be little justification for that response.

In 1964 there really was cause for concern, just as there is today. But the problem has never been that Japan (or some other country) is doing better than the United States is. The problem is that we are not doing nearly as good a job as possible to help all of our children learn and understand enough mathematics to lead productive and fulfilling lives in a modern society. We have always failed to teach mathematics so that people would be willing and able to use it effectively. Despite that failure, both individuals and society have usually managed to muddle through. Muddling through without the appropriate attitudes and abilities in mathematics has become, and will continue to become, more difficult—for both individuals and society.

Reforms of the past were driven by individuals and organizations, but tended to die when advocates reduced their efforts. Today we have a different situation. Strong individual and organizational voices are advocating reform, but the real motivation for reform is a change in society itself. Never before has a change in technology made knowledge and understanding of mathematics so important to so many people. Never before has a change in technology made the kind of mathematics most people have been learning so obsolete. The technological revolution will not go away. We will not collect and destroy all calculators and computers on some day in the future. The reformers may die, but the reforms now taking place will continue to live. Those who fail to benefit from these reforms will live less full and less productive lives than those who benefit from the reforms. Those societies that prepare people well for a technological future will become better places to live. Those that don't will wither.

In this book, I review recent, widely supported recommendations; provide specific examples; and make suggestions for change. In addition, I consider changes in technology and the effects those changes are having on both the goals of mathematics education and the means of delivering mathematics education. Finally, I present some suggestions for ways in which concerned educators can foster constructive changes in the teaching of mathematics.

This book does not represent an attempt to provide all the information needed to make and support all necessary change. Rather, I hope to help the generalist understand where we are, why we should change, where we should be going, and how to get there. I have not tried to cite all related scholarly research, but I have included some references for further reading. Three sets of "standards" produced by the National Council of Teachers of Mathematics [*Standards for Selection and Implementation of Instructional Materials* (1984), *Curriculum and Evaluation Standards for School Mathematics* (1989), and *Professional Standards for Teaching Mathematics* (1990)] are likely to be particularly helpful for those contemplating change.

Why Is Change Needed?

In many respects, U.S. education is doing more for more children with fewer resources and less appreciation than has ever been true in the history of any country. Certainly, our job would be easier if we were not expected to solve all the problems generated by the breakdown of the American family, the drug culture, the short attention spans of children in a multimedia age, and the many other problems of society.

But we don't have that choice. These problems will continue to be with us. Many people have spoken eloquently and constructively (unfortunately, with little impact) about these problems; but I address them here only insofar as they provide the larger context within which we must teach mathematics.

Has mathematics education become much worse in the recent past? Anyone who has read the many media reports about mathematics education, the several international studies, or the analyses of National Assessment of Educational Progress (NAEP) results could be excused for assuming that there has been a disastrous deterioration in the teaching of mathematics. In fact, that's not true. A careful analysis of these various reports and other available information shows that the teaching of mathematics today is probably not much worse, and not much better, than it has been at any time in the past 50 years.

Throughout history, most human beings have been unwilling or unable to do simple mathematics. Even in those few times and countries where large numbers of people have been taught some mathematics, their general belief has been that mathematics is akin to mysticism—formulas and procedures handed down on stone tablets, to be memorized and used when necessary, but never to be understood by mere mortals. Of course, the idea that normal human beings might create mathematics to help solve the real-world problems around them would never occur to most people, even if they had successfully studied mathematics for ten years or more.

The inability of most people to think mathematically has been an unfortunate negative factor in civilization's progress over the years. Recent advances in technology, however, and a world that is becoming continually more complex and quantitative, show us that mathematical thinking is becoming ever more important. Furthermore, the kind of mathematics that people will need in an age of computers and calculators is different from the mathematics needed in the past. Simple symbol manipulation can be done effectively by machines, but higher-order thinking skills and the ability to communicate intelligently about mathematical situations are still uniquely human skills.

That does not mean we no longer need the lower-order skills. Knowing the addition, subtraction, multiplication, and division facts "by heart" (or by mind) is at least as important as it ever was, and many other lower-order skills are still essential so that we can concentrate on the higher-order skills. But people of the future will need more, better, and different mathematical skills. They must be able to relate their mathematical skills to the world around them and use their skills to help solve problems that are of importance to themselves and to the world.

Change in the teaching of mathematics is needed, not because it has recently deteriorated, nor even because it has always been bad (though it probably has been), but, rather, because the world is changing. The people who are going to solve the problems of the present and future—or even understand and evaluate those problems and solutions—must have a far better grasp of mathematics than most people have at present, or have ever had in the past.

Most technologically advanced societies have recognized the need for better mathematics education and are devoting enormous national, local, and individual resources to the task of improving the teaching of mathematics. This has resulted in substantial disparities among the international mathematics scores of children, and the United States looks less adequate than most of the others.

If you talk to educators, parents, or even children in these other countries, however, you are struck with the fact that they are, in general, *not* satisfied with their mathematics education. But in the United States, where children are demonstrably learning less mathematics and learning it less well than in other industrialized societies, teachers, parents, and children are convinced that they are doing well in mathematics. This is borne out by every recent international study, including reports by Crosswhite (1984) and Stevenson (1986). Stevenson argues that the main differences in mathematics education between the United States and Japan are that the Japanese curriculum is much more challenging, Japanese children spend far more time on task in school, Japanese children work harder, and a Japanese child who is having trouble learning is expected to work *harder*, whereas in the United States similar children would be excused because the mathematics is too difficult for them.

We can probably learn to follow, rather than lead, in the solution of the world's problems. We can probably live quite comfortably as a second-rate economic and military nation. But some of our children may wish to participate fully in the exciting life of the next millennium that is about to burst upon us. They will need a much better mathematics education than is generally available in the United States. Indeed, they will need a better education in mathematics than is generally available in any country. We should give them an opportunity to learn enough to participate fully in their world of the future.

We can and must do better. The goal of education—the goal of good teachers everywhere—is to educate people who will go beyond where we have gone. The people who believe "if it was good enough for me and my parents, then it is good enough for my children and their children" will be left standing in the dust as the world moves by them.

Can We Improve?

Recently a mathematician said to me: "You guys have been working on the problems of mathematics education for more than 30 years now. Why don't you have them all solved?" He was serious.

I wondered briefly why mathematicians who have been working on the problem called "Fermat's Last Theorem" for more than 300 years have not yet solved it. I also wondered how well my colleague would do solving problems in mathematics education where success depends on all sorts of variables that change from day to day and year to year—so that a "solution" for the 1930s is often not even pertinent to the 1990s; where a "solution" for New York City is thought laughable in Bisbee, Arizona (and vice versa); and where political considerations are likely to play a crucial role in determining the success of (and even the opportunity to try) new curriculums and methods.

In education, we do not march steadily and unhesitatingly forward. We repeat not only the errors of the past, but also the successes—usually without knowing we are repeating ourselves. But worse, we regularly find that the procedures that failed at some time in the past are successful at a later date, and the procedures that were successful no longer succeed.

As we evaluate some particular movement in education, we should try to remember that education is not a science. It is a profession that combines both art and science. We can ask reasonable questions about attempts to reform education, and we can use both art and science to help answer those questions. Were the goals reasonable in light of the needs of society? Were the goals achieved? What effects, if any, linger after the movement has died? What can we learn from the movement that will help us do better next time? In this spirit, let us examine the "new mathematics" of the 1950s and 1960s.

Why Didn't the "New Math" Work?

The new math movement had both positive and negative effects. Overall, there was probably a more positive than negative influence. As we continue to try to improve the teaching of mathematics, we should remember which were which, and try to avoid the negative aspects while pursuing positive goals. Many of the negative effects of the new math activities could probably have been avoided if people who were leading the movement had spent more time in real classrooms working with real children and teachers, and had listened intently to those children and teachers. For example, the undue emphasis on logical rigor was totally inappropriate for children and even for adults. On the other hand, integration of the various topics in the mathematics curriculum and an attempt to show more applications of mathematics to real life and other academic pursuits were highly desirable, as was the attempt to modernize school mathematics by bringing some of the mathematics created in the past 350 years into the school curriculum.

The "acceleration" emphasis of the new math movement was a mixed bag. Certainly our children had both the need and ability to learn more and better mathematics, but dividing 7th grade children into "good students" who can easily do two years of mathematics in one year and "the others" was unfortunate, at best. James Flanders analyzed three of the better selling textbook series in 1987 to see how much new material was introduced in each grade. The results were scandalous. Giving credit to a book for a "new" page if any new material appears on that page, Flanders concluded that the average percentages of "new" pages for each grade from kindergarten through grade 9 for the three series are:

Grade	Percentage
K	100
1	75
2	40
3	60
4	45
5	50
6	38
7	35
8	30
9	90

What could possibly justify this "dumbing down" of the mathematics curriculum, particularly for grades 6, 7, and 8? How could we hope that children who can learn only 30 percent new material in grade 8 could learn three times that much in grade 9?

Normal children in most other industrialized countries learn far more mathematics in grades K–8 than is typical in the United States, with no apparent ill effects. The variations shown by Flanders make it clear that children can learn more in grades 2 and 4, as well as grades 6–8. We must begin to encourage children and teachers to fulfill their potential and expect reasonably challenging curricular materials for all grades.

By dividing children along imagined ability lines at the start of 7th grade and allowing the "good" ones to pursue a reasonably challenging mathematics curriculum while relegating the others to an intellectual wasteland, the acceleration of the 1960s institutionalized a vacuous curriculum for grades 7 and 8. As long as parents and teachers of the "good" children (read "middle class") are happy, there is little chance of correcting the real problem.

There were both good and bad aspects of the new math movement, but the overformalism and the lack of any obvious connection to the real world strengthened opponents of the movement when nostalgic, unenlightened pedants took us squarely back into the 19th century with the back-to-basics movement.

Lately, conventional wisdom has claimed that the apparent failure of the new math occurred because change was expected too quickly. I believe this is wrong. There were substantial positive changes that occurred very quickly, many of which have remained with us. If education has any hope of preparing today's children for tomorrow's world, change must come quickly. The major failure of the new math was in direction, not velocity. Too many of the goals simply were wrongheaded. That provided ammunition for those who were opposed to all changes, including the good ones.

References and Further Resources

Crosswhite, F.J. (1984). *Second Study of Mathematics*. Champaign, Ill.: International Association for the Evaluation of Educational Achievement, U.S. National Coordinating Center.

Dossey, J.A. (1989). "Transforming Mathematics Education." *Educational Leadership* 47: 22-24.

Dossey, J.A., I.V.S. Mullis, M.M. Lindquist, and D.L. Chambers. (1988). *Mathematics: Are We Measuring Up?* Princeton, N.J.: Educational Testing Service.

Driscoll, M. (January 1988). "Transforming the 'Underachieving' Math Curriculum." *ASCD Curriculum Update*.

Flanders, J.R. (1987). "How Much of the Content in Mathematics Textbooks Is New?" *Arithmetic Teacher* 35: 18-23.

McKnight, C.C., F.J. Crosswhite, J.A. Dossey, E. Kifer, J.O. Swafford, K.J. Travers, and T.J. Cooney. (1987). *The Underachieving Curriculum: Assessing U.S. School Mathematics from an International Perspective*. Champaign, Ill.: Stipes.

National Council of Teachers of Mathematics. (1984). *Standards for Selection and Implementation of Instructional Materials*. Reston, Va.: NCTM.

National Council of Teachers of Mathematics. (1989). *Curriculum and Evaluation Standards for School Mathematics*. Reston, Va.: NCTM.

National Council of Teachers of Mathematics. (1990). *Professional Standards for Teaching Mathematics*. Reston, Va.: NCTM.

Stevenson, H.W., ed. (1986). *Child Development and Education in Japan*. New York: W.H. Freeman.

Recent Recommendations of Professional Groups

Even before new math materials were widely adopted, some people had begun to raise serious questions about various aspects of the "revolution" and to propose somewhat different goals and implementation procedures for future reforms in school mathematics. The similarity among such recommendations over the past 15 years is as striking as the rapidly growing number of people who subscribe to essentially the same views.

Different Skills Emphasized

Though the various recommendations differ in some details, most of the position papers produced by major textbook-adoption states, leading professional organizations, and individuals involved in mathematics education since 1975 have recommended that schoolchildren become proficient in the following basic skills:

1. Problem solving
2. Communication skills related to mathematics
3. Integration of topics within mathematics
4. Relating mathematics to other subjects and to the learner's real world (closely associated with number 1)
5. Understanding and using functions, relations, and patterns
6. Probability and statistics
7. Approximation and estimation
8. Using number sense and number systems effectively
9. Computational ability—both written and mental computation—with rational and some irrational numbers
10. Two- and three-dimensional geometry, including both synthetic and algebraic arguments in geometry
11. Understanding and using mathematical structures
12. Measurement
13. Algebra

14. Trigonometry
15. Discrete mathematics
16. Intuitive foundations of calculus
17. Using technology (calculators, computers, etc.) to help solve mathematical problems

Beyond this, all groups have recommended that mathematics be taught in such a way that people not only will be able to use mathematics to solve problems, but also will *want* to use mathematics, and will think of mathematics as a friendly, useful tool, rather than a nemesis to be avoided at all costs.

Many of these recommendations reflect appropriate goals for teaching mathematics at any time and in any place. Some, however, suggest both content and methods that are quite different from most mathematics teaching of the past. Much of the rest of this book is devoted to discussing some of the more striking differences and how teachers and supervisors can facilitate the mathematics education of children to prepare them better for life in the 21st century.

To treat each recommendation individually with any degree of completeness would be both inappropriate and impossible in this book. Certain principles and procedures, however, will help educators achieve the general goals. These include: decompartmentalizing mathematics, relating mathematics to other subjects and the real world, improving problem-solving skills, enhancing communication about mathematics, and fostering more positive attitudes toward mathematics. Other pertinent issues include procedures for introducing and doing mathematics that will enhance students' ability and desire to use it intelligently, the effects of curricular material and testing on learning, the choice of appropriate materials, and the selection and support of teachers.

Moreover, some of the recommendations are either so important, or suggest activities that are so different from traditional practice, that we will spend some time examining them.

Four Steps to Better Mathematics Education

There are four important steps that children should follow to learn mathematics and to be willing and able to use it effectively to solve problems of all kinds: (1) derive the mathematics from their own reality, (2) discover and use the power of abstract thought, (3) practice, and (4) apply the mathematics to something that is of interest to them.

Derive Mathematics from the Learner's Reality

Max Beberman was one of the great mathematics educators of this century. One day his seven-year-old daughter came home from school with several subtraction problems to do. Her father noticed she was getting strange answers. For example, to the problem: 32 − 18, she would get an answer of 26. He watched to see what she was doing. First she lined up the numbers with tens above tens and units above units. Then she subtracted 1 from 3, getting 2, which she wrote in the tens column of the answer space. Next, since she couldn't subtract 8 from 2, she subtracted 2 from 8 and wrote the resulting 6 in the units column of the answer space. Everyone who has taught 2nd grade—and many people who have taught 3rd, 4th, and even 5th grade—has seen this phenomenon.

$$\begin{array}{r} 32 \\ -18 \\ \hline 26 \end{array}$$

Professor Beberman decided to correct the situation using procedures known to excellent teachers for decades. He got out a bunch of ice cream sticks and had his daughter group them with rubber bands into bunches of ten each until she had 32 sticks (three groups of ten and two more) (see Figure 2.1). He then asked her to take 18 sticks away.

Figure 2.1

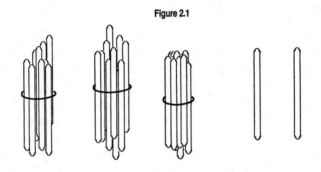

She removed one bunch of ten and examined the situation for a while. "Is it all right if I take a rubber band off one bunch of ten?" she asked. Her father's beaming face made it clear she had hit on the right solution. She proceeded to remove the rubber band and take away eight more sticks.

"How many sticks do you have left?" asked Professor Beberman. "Ten and 4 more—14," she responded.

"Now do you see that 32 minus 18 is 14, not 26?" prompted her father.

She thought about this interesting turn of events for a moment or two and responded: "With ice cream sticks the answer is 14. In school the answer is 26."

This story typifies a serious problem with mathematics education. Intelligent students decide early in their schooling that the mathematics they learn in school has nothing to do with their real world outside school. They see no conflict between getting the answer "26" to a problem done by school methods and getting the answer "14" when the same problem is done with real objects. They believe the mathematics of school has nothing to do with their real world. They do mathematics in school to please the teacher or their parents, or they do it to get done with it so people will stop bothering them, or for some other reason—but almost never do they do such problems because they want to know the answer. And almost never do they have any reason to believe that the problem, or its solution, has anything whatsoever to do with their real world.

All of mathematics can, and should, be derived from something that is real to the learner. For very young children, that reality usually is something physical such as ice cream sticks, fingers, chairs, people, distances walked, heights of friends and relatives, and so on. For older students, the reality may be the natural or social sciences, puzzles, or even some part of mathematics itself. For example, arithmetic, which is quite abstract to a six-year-old, may be just the reality needed to help an older student understand algebra.

There is an unfortunate belief among many teachers of mathematics that manipulative materials, and other references to reality, are designed for the "slower" student, but that the "good" students don't really need that sort of thing. Fairly bright children can often learn to manipulate mathematical symbols reasonably efficiently, with little or no understanding, in a way that satisfies teachers. But such symbol manipulation, without physical or other real referents, is precisely what leads these "good" students to believe that mathematics has nothing to do with their real worlds. As long as they are successful at the symbol manipulation, such students often find the success sufficient reward in itself. But as soon as they run into difficulties, the more intelligent among them begin to suspect that this abstract game that is so thoroughly removed from reality has little place in their lives and should be set aside for more interesting and useful activities.

Sooner or later in their formal education, most people begin to believe that mathematics is a masochistic way to make people work very hard when they would otherwise not need to work at all. Nothing could be further from the truth. In many respects, mathematics is a lazy person's activity. It is a way to solve problems more easily than they could be solved otherwise, or a way to solve problems that simply could not be solved by any other means. By deriving mathematics from the learner's reality and by constantly applying that mathematics back to real situations in which the learner is interested, we can help students to understand mathematics better and to see it as a useful, powerful, and even beautiful tool that helps them solve their problems and helps them understand the world around them better.

The Power of Mathematics Is in Its Abstractness

Deriving mathematics from the learner's reality is an important part of teaching mathematics in a meaningful way. But the fact that the power and beauty of mathematics is in its abstractness is equally important and must not be neglected. Some people, having understood the first point, take it so seriously that they completely lose sight of the second.

To give a simple example, suppose we had different kinds of numbers for counting people, for counting chairs, for counting dogs, and so on. Then, if we discovered that 2 people plus 3 people is always 5 people, we might still not know how many a pair of chairs plus a triple of chairs is because we couldn't generalize our results for people to other objects. Nor would we know the corresponding answers for dogs, cats, or other things. The fact that numbers and other objects of mathematics are essentially independent of the specific reality from which they were derived is what gives mathematics its power. Having learned that $3 + 5 = 8$, we can be sure that 8 will be the total when we have brought together sets of 3 dogs and 5 dogs, 3 chairs and 5 chairs, 3 widgets and 5 widgets, or whatever. In a similar way, the other objects of mathematics are abstractions about which we can reason quite independently of any specific reality—and having reasoned abstractly, we can be sure that our results apply to any real situation for which the original assumptions applied.

The apparent conflict between the abstractness of mathematics and its close connection to the real world gives rise to many problems in mathematics education. Some teachers, having understood the abstract nature of mathematics, believe that it should be taught in a totally abstract way with no reliance on its connection to the real world. Others, having realized that mathematics is motivated by situations in the real world, is closely allied to the real world, and can be useful in solving real-world problems, believe that all mathematics should be learned, taught, and discussed in connection with some physical referent—never relying on totally abstract symbols. Either position, carried to an extreme, will defeat the learning of mathematics.

To help young children learn mathematics with an understanding and appreciation of its role in the real world, and at the same time let them see the power of its abstractness, we should use several different physical models for the same mathematical concept. We should help pupils see that mathematical operations turn out the same no matter what the physical referent. For example, after using the ice cream sticks and rubber bands, Max Beberman could have also used base ten blocks, or three children each holding up 10 fingers and a fourth child holding up 2 fingers, or three decimeters (each with 10 centimeters marked) and 2 more centimeters, or any one of many other representations of 32 to help convince his daughter that when 18 is subtracted from 32 the answer always comes out 14. Several such models might have helped her understand the nature of the mathematics better.

Having seen how mathematics can be derived from the real world, children must also learn to work with mathematics in its pure abstract form. Children should realize that when they reason correctly with the pure abstractions, the results will be applicable to any reality for which the original mathematics was appropriate. They should also learn that reasoning with the mathematical symbols is usually both easier and more efficient than working with the physical referents. Imagine doing every addition or subtraction problem with ice cream sticks or other similar objects. Just lugging the stuff around with you would be more trouble than most people would be willing to accept. Then imagine trying to do the same for multiplication, division, and so on.

Almost any good procedure can become counterproductive when carried to extremes. When teaching mathematics, we should certainly show the connection between the real world and the mathematics; but we must also remember that the power and beauty of the mathematics is in its abstractness. We should not try to obscure the abstract nature of mathematics by forcing children (or encouraging them) to use physical objects after sufficient initial connection has been made. And we certainly should not limit children to only one physical referent—no matter how good that referent may seem, or how persuasive the salesperson is who advocates the use of that material.

Practice Makes Perfect

From time to time, some educators subscribe to the strange notion that if a student really understands a concept or skill, that student will always remember the concept and will be able to use the skill without having practiced it. Those of us who have spent our lives teaching children know better. We have all had the experience of watching a person who completely understands a concept one minute forget it the next. We have watched students perform a task perfectly one day and then seen them have no idea of how to get started on the same task, or an equivalent one, the next day. From such experiences we have concluded that if we want students to be really good at a particular skill, or if we want them to really remember and understand a concept, we must arrange for them to practice.

Practice need not be unpleasant. Indeed, practice is far more effective if it is pleasant. E.L. Thorndike is often thought of as the patron saint of behaviorism—including the idea that pupils should practice correct responses and thus reinforce them. But Thorndike did not advocate the unpleasant, punitive kind of practice that has often been associated with his name. He believed learners should be rewarded for correct responses, not punished for incorrect ones. He never suggested that the practice itself should be so unpleasant as to seem like punishment. The "drill-and-kill" approach, epitomized by the distribution of dozens of dreary ditto sheets or textbooks that are little more than hardbound ditto sheets, may have the advantage of keeping the pupils quiet while the teacher does something else, but it is generally not an effective means of providing practice.

There are many ways to practice skills so that the learner has positive associations with them and is likely to continue to be willing and able to use them even after required schooling. Carefully conceived educational games provide one of the most effective methods of practice. Practice with people or even machines or cards that give positive, immediate feedback can also be effective for some learners. Projects or activities that require mathematical skills and concepts to help students reach some goal they wish to reach can provide excellent practice and can have the additional advantage of encouraging learners to see mathematics as useful.

These and various other methods of effective practice generally require more work by the textbook authors and teachers than multiple pages of dreary "exercises," but the results are dramatically better. Most teachers think the extra effort is well spent.

Provide Realistic Applications

"5 lights on. 3 lights off. How many lights on?"

This typical word problem from a 1st grade mathematics textbook is a classic case of a bit of nonsense masquerading as an "application" of mathematics. What's wrong with this "problem"?

First, the failure of the authors to write in English sentences is pernicious. One of the important goals of teaching mathematics to human beings should be to help them communicate with other people about mathematics. No normal person communicates in the kind of "telegraphese" used in this and similar "word problems." If children are too young to read correct English (or whatever the language of instruction is) they should be introduced to problems orally or through physical activity until they are able to read.

Problems like this one are a result of the blind, unthinking use of reading-level formulas by textbook-adoption groups. Textbook publishers have discovered they can fool the formulas by substituting short non-sentences for real sentences. Never mind that such procedures are counter to the goals of both reading and mathematics instruction. Never mind that children have a harder time reading these short, incomprehensible sentences than longer, correct sentences. If the calculated reading level can be kept low enough, the textbook will be adopted.

Second, to suppose the child has read this problem and understood it when the child answers "2" is naive. Any normal adult who reads the problem will answer "5," unless the adult has been brainwashed by years of schooling into believing the answer is never provided in the problem and therefore the reader is expected to perform some mathematical operation to arrive at the desired answer. A child will get the "correct" answer to such a problem by noticing how other problems on the page are done, then doing this one the same way. Or perhaps the child will recall which operation has been taught most recently and use that. Either method would produce the expected answer to this problem because of the way the textbook was designed. Both methods show a certain ability to solve the problem of how to get along in a somewhat irrational adult world. Neither is what we ordinarily think of as mathematical problem solving or applying mathematics to real situations.

Third, suppose the child is sufficiently sophisticated to read the problem, conclude that the author doesn't know what a verb is and that the author meant to say: "There were five lights on in a room. Somebody turned off three of those lights. After that, how many lights were still on?" Surely this sophisticated child would also realize that subtraction is not an efficient method of solving the problem. To find out how many lights were on in the beginning, somebody had to count all five of them. Then somebody had to count the three that were turned off. Finally, the child is expected to do a subtraction problem with those two pieces of data. Why not just count the two that are left on?

The third objection may seem like a petty point, and one that children are not likely to notice, but it is really terribly important. If students are regularly required to solve word problems by complex mathematical techniques that are less efficient than other procedures known to them, they will develop a feeling that mathematics is a way to make them work when they would not otherwise have to work.

When I was growing up in Wisconsin, I heard a story about two travelers on a train. One looked out the window and commented on the many cattle in a farmyard. The second traveler glanced out the window and said, "Yes, there are 187 of them." The first asked in awe, "How did you get that number so fast?" To which the second responded, "Simple. I'm a mathematician. I counted the legs and divided by 4."

Unfortunately, this apocryphal story accurately represents the views of many people about mathematics and mathematicians. Years of formal schooling have left people with the idea that mathematics is a mass of memorized mental tricks that are to be used in an unthinking and inefficient manner to make problem solvers work harder than they would otherwise have to work.

Traditional word problems may occasionally give people an idea of reasonably realistic applications of mathematics; and they are easier to produce, to administer, and to use for evaluating students than are real applications that are interesting and worthwhile. But even when they are good, such word problems are a poor substitute for applications that would help the learners see the usefulness and beauty of mathematics. When they are artificial "make-work" problems, such as the "lights on" example, they do more harm than good.

Situations taken from pupils' real lives often provide the best applications: What is the most efficient way to get to school? How early do you have to leave home to be reasonably sure you'll get to school on time? Which grocery store offers the best buys? Should quality, convenience, and other matters influence your decision? How much money will I need in order to buy a bicycle? How can I manage to get that much money? If I smoke three packs of cigarettes a day and also fly 1,000 miles on airplanes every day, am I more likely to die from the effects of smoking or from injuries sustained in an airplane crash?

Students can also create their own problems from news articles, from their daily activities, or even from their imaginations. Chapter 4 of this book discusses procedures for using newspapers to help children formulate their own problems.

Another kind of application that is real to children, but may not seem entirely real to all adults, is the intelligent playing of certain games. A good mathematics game ordinarily should provide practice in some particular skill, but it also can have good problems imbedded in it so that players can improve their strategies by thinking mathematically. Children should be able to play the game without solving the problems, but they should be able to play it better if they do solve those problems. This is analogous to real life in which people can, and do, get by without using mathematics, but can live and understand better if they are able to think mathematically. Several such games are described later in this book (see Chapters 3 and 4).

The activities suggested here require more work on the part of authors, teachers, and pupils. But activities that are meaningful to the students are more likely to be remembered and more likely to leave the learners with a feeling that mathematics is useful and worth learning. Surely that is better than having students believe that mathematics is a subject they are required to learn to satisfy other people, and that it should be put out of their minds as soon as possible.

References and Further Resources

Conference Board of the Mathematical Sciences. (1983). "The Mathematical Sciences Curriculum K–12: What Is Still Fundamental and What Is Not." In *Educating Americans for the 21st Century. Vol. 2. Source Materials,* edited by the National Science Board Commission on Precollege Education in Mathematics, Science and Technology. Washington, D.C.: National Science Foundation.

Conference Board of the Mathematical Sciences, National Advisory Committee on Mathematical Education. (1975). *Overview and Analysis of School Mathematics, Grades K–12.* Washington, D.C.: Conference Board of the Mathematical Sciences.

Dossey, J.A. (1989). "Transforming Mathematics Education." *Educational Leadership* 47, 4: 22–24.

Dossey, J.A., I.V.S. Mullis, M.M. Lindquist, and D.L. Chambers. (1988). *Mathematics: Are We Measuring Up?* Princeton, N.J.: Educational Testing Service.

Driscoll, M. (January 1988). "Transforming the 'Underachieving' Math Curriculum." *ASCD Curriculum Update.*

Kline, M. (1966). "A Proposal for the High School Mathematics Curriculum." *The Mathematics Teacher* 49: 322–330.

Kline, M. (1973). *Why Johnny Can't Add: The Failure of the New Math.* New York: St. Martin's Press.

Mathematical Sciences Education Board. (1989). *Everybody Counts.* Washington, D.C.: National Academy Press.

McKnight, C.C. et al. (1987). *The Underachieving Curriculum: Assessing U.S. School Mathematics from an International Perspective.* Champaign, Ill.: Stipes.

National Council of Supervisors of Mathematics. (1976). *Position Paper on Basic Mathematical Skills.* National Council of Supervisors of Mathematics. (Available from the

Council, c/o Ross Taylor, Minneapolis Public Schools, 807 Broadway N.E., Minneapolis, MN 55413.)

National Council of Teachers of Mathematics. (1980). *An Agenda for Action: Recommendations for School Mathematics of the 1980s*. Reston, Va.: NCTM.

National Council of Teachers of Mathematics. (1989). *Curriculum and Evaluation Standards for School Mathematics*. Reston, Va.: NCTM.

National Institute of Education. (1975). *Conference on Basic Mathematical Skills and Learning, Euclid, Ohio*. Washington, D.C.: National Institute of Education.

National Science Board Commission on Precollege Education in Mathematics, Science, and Technology. (1983). *Educating Americans for the 21st Century*. 2 vols. Washington, D.C.: National Science Foundation.

Willoughby, S.S. (1967). "Revolution, Rigor and Rigor Mortis." *Mathematics Teacher* 60: 105–108.

Willoughby, S.S. (1981). *Teaching Mathematics: What is Basic?* (Occasional Paper #31). Washington, D.C.: Council for Basic Education.

Willoughby, S.S., C. Bereiter, P. Hilton, and J.H. Rubinstein. (1981, 1985, 1987). *Real Math, K-8 Textbook Series*. La Salle, Ill.: Open Court.

Implementing Recommendations

The four steps suggested in Chapter 2 (start with learner's reality, abstract, practice, apply) are more meaningful if viewed in a context. The principles are quite general and can be applied to almost any topic in elementary school, in high school, and in graduate school mathematics. This chapter presents examples involving early elementary school computation to keep the mathematical content as unoppressive and familiar as possible. Examples from topics that have recently been incorporated into the curriculum are easy to provide, but it is important for educators to realize that we can also improve substantially the way we approach the more traditional topics.

Example 1: Single-Digit Addition

This rather extensive example demonstrates methods of achieving many of the goals recommended by various professional organizations, such as improved problem solving and communication and greater integration of mathematical topics horizontally and vertically. However, to avoid making the example even longer, many steps have been condensed or omitted.

Learning About the Learner's Reality

Before beginning any new topic in mathematics, the teacher should check to see that pupils have the necessary background. To develop a concept from the learner's reality, we must first establish what the learner's reality is.

In the case of addition, the essential concept has to do with how many objects are in a set of objects that results from combining two other sets. To an average adult, the underlying concepts seem simple, and there appears to be no doubt that a normal six-year-old would be able to understand everything necessary. But suppose the child doesn't believe that the number of objects in a set remains the same when they are rearranged.

Then, when two sets are brought together, there is no reason for the child to expect that the number of objects in the combined set is related in any rational way to the numbers of objects in the two original sets.

Piaget and others reported experiments with young children that seem to indicate that many children do not "conserve number." That is, many young children (under the age of five years) do not seem to believe the number of objects in a set remains the same when the objects are rearranged. Conservation of number (or area or volume or other things) is important, but the essential underlying general principle is that teachers must make a serious effort to understand what is going on in the minds of their pupils. This requires asking some searching questions; more important, it requires listening very carefully with a mind that is as open as we can possibly make it. Sometimes we will find that we are totally unable to fathom what is going on in the minds of our pupils, but we must always make the effort. In a class of 25 or 30 children, this initial evaluation is anything but easy.

Experiences from the Learner's Reality

Once a child conserves number and can count (let's say to 20) we can start concrete experiences with addition:

Here are four coins. We put three more coins with them. How many coins are there now?

Repeat this with numerous other kinds of objects, such as crayons, pencils, sticks, fingers, children, desks, and so forth. Include pairs of problems, such as:

A. Here are five sticks. I put two more with them. How many are there now?
B. Here are two sticks. I put five more with them. How many are there now?

If the child really conserves number and has any inclination at all to avoid work by thinking (an essential characteristic of mathematically inclined individuals), the second problem should not require counting. A child who gets the second problem of a pair like this without any hesitation probably understands the important principle that the order in which we add two numbers makes no difference in the answer. That is, for any two numbers A and B, $A + B = B + A$. This is called the *commutative principle of addition*. A good understanding of the commutative principle will reduce by almost half the number of addition facts that must be learned, and is, therefore, well worth knowing and stating explicitly (but not necessarily with the formal name).

In their early, concrete addition activities, children will count the number of objects in each addend and then count to find out how many are in the sum. To add $5 + 4$, the child will count out five objects, then count out four objects, then put the two bunches together and count the total number (nine). Is there a more efficient method?

For a real problem, if you know there are five raspberries and you want to know how many there would be if you added four more, you'd "count on" from five as you added the four: "six (one berry in), seven (second berry in), eight (third berry in), nine (fourth berry in), so the total is nine." But if we don't actually have the berries, and want to represent them with something we do have, we will have to count out each of the sets unless we can find a more efficient method.

Physical Representations and Finger Phobia

A physical manipulative material that is of great value in mathematics and is always available for most children is fingers. Many elementary school teachers have built up a sort of phobia about fingers because so many children seem to misuse them in later grades. If a child is still counting out addition and subtraction problems on fingers in the 5th or 6th grade, then that child is wasting valuable time and thought doing something that should be automatic. That doesn't mean the child is evil, nor should we put mittens on the fingers to make the child do it mentally. Rather, in the early grades, we should try to teach so that most children will become very efficient at simple operations and can spend their time and thought on more advanced ideas.

It does not follow that children should not use their fingers in early grades. Quite the contrary. If we teach children to use their fingers intelligently in the early grades, they should be more able to get along without using them later. Furthermore, if we teach children to use their fingers at the age of six, then when they are eight they will think of this as a juvenile activity and try to avoid it. On the other hand, if we let our finger phobia dominate in the early years, children will decide that using fingers is a subversive activity, and will continue to do it surreptitiously as long as they can. We must not become fanatic about fingers. Our goal is to have children know and use the addition and subtraction facts efficiently when they are eight and older. If we can achieve this by teaching them to use fingers at the age of six, we should do so.

An efficient way to teach young children to add involves always using the same fingers to stand for a given number. Thus (for example), the thumb up on the right hand is always used to represent one. The thumb and forefinger on the right hand always represents two. The thumb, forefinger, and middle finger represents three. All the fingers on the right hand except the little finger are used to represent four, and all the fingers on the right hand are used to represent five. Continuing in the same way, the fingers on the right hand combined with the thumb of the left hand represent six, the fingers on the right hand and the thumb and forefinger of the left hand represent seven, and so on.

Once we have learned these "finger sets," if we want to know what 6 + 3 is, we can start by putting up the finger set for six without the need to count. Then we can count on as we put up more fingers: "one (left forefinger goes up), two (left middle finger goes up), three (left ring finger goes up)." Now, we know, without counting, that we have the finger set for nine, so, we know that 6 + 3 = 9.

We have solved our problem by counting only one set (the set of three) rather than having to count all three sets (six, three, and nine).

Notice that if we really understand the commutative law for addition we can always arrange to count the smaller of two unequal addends. If the problem is 3 + 6, we can simply turn it around and make it 6 + 3.

Having taught children to use finger sets to help learn addition facts, we certainly should feel obliged to show them how to move beyond finger sets. We might try something called "statue arithmetic," in which children are allowed to do the previous problem by putting up a finger set of six, but are then required to be statues (they can't move their fingers). Then, they imagine what set they would have if they counted out three more. Finally, you can ask them to imagine the six-set, then imagine adding the three-set, and then visualize the answer.

A great deal of physical experience should accompany the learning of addition facts. Such activities might include being told there are five counters (or pebbles, coins, sticks, or other objects) in a can, then watching (and presumably counting) as somebody puts three more in, then deciding how many counters there are in the can, and other similar experiments.

To relate counting and addition to geometry and measurement, we might count how many of a child's feet (heel to toe) it takes to walk the length of a table. We might then put two tables together and try to predict how many of the child's feet will be required for the double length. We could do the same with Cuisenaire rods, or with plastic or cardboard number strips (ranging in size from one unit to ten units—an appropriate unit would be about two centimeters or one inch in length), or similar objects.

The drive strips (or "holy paper") torn off the sides of computer printout paper after printing can also be used for measuring—the number of strips plus the number of holes provide a reasonably precise measurement. Conversion between numbers of strips and numbers of holes seems more concrete to young children than conversion between feet and inches or between meters and centimeters.

After a great deal of this sort of activity and, if necessary, specific related experiments, children should begin to realize that if 3 apples plus 4 apples is 7 apples, then 3 people plus 4 people will be 7 people, and 3 pencils plus 4 pencils is 7 pencils, and so on. This may seem like a trivial insight to an adult, but it is the very essence of mathematics. We have started from multiple physical models and abstracted general principles that will work for all those models and infinitely many more models. We can now use abstract arithmetic that will work for all these situations. That is the true power of mathematics.

By the time children are ready to learn the addition facts in a formal way, they should have become quite good at adding one, two, and three to any number by simply counting on. They are also likely to have learned the 5's facts if they have been using finger sets as described previously, since all the numbers up to ten are shown with all five fingers of the right hand up plus some more on the left hand, and those on the left hand correspond to the appropriate number on the right hand. For sums beyond 5 + 5, consider how much greater than five the second addend is. For example, in 5 + 7, the seven will be thought of as five on the right hand and two more. Put the two fives together to make ten, and the answer is ten and two more.

Base Ten

Before continuing with this addition example, it is worth interjecting a comment about base ten. The numeration system we use is based on the number ten. All the standard procedures (or algorithms) we use to add, subtract, multiply, divide, and do other operations depend on the fact that we are using a base ten numeration system. Because of this, if children are explicitly aware of the fact that their numeration system is base ten, they will generally find it much easier to understand and learn the various algorithms.

Therefore, most early work with numbers greater than ten should emphasize base ten concepts. Thus, when teaching children to count above ten, names such as "ten and one, ten and two, ten and three, . . . ten and nine, two tens, two tens and one," are more appropriate than the traditional names of "eleven, twelve, thirteen, . . . nineteen, twenty, twenty-one." Children will have heard the traditional names and should not be prevented from using them, but they should also be encouraged to use base ten names. Notice that, with the exception of "eleven" and "twelve," the usual words for numbers all have the base ten idea inherent in them. For example, "fourteen" means four and ten (or "teen"), "sixty" comes from six tens ("ty" stands for tens here), and so on. Some of the words (like "thirteen") seem a bit far from the original, but a little imagination makes the connection seem reasonable.

With this interjection in mind, ten and two is a perfectly reasonable answer to the question of what is 7 plus 5. We also realize that all the 10's facts are automatic: 10 plus 7 is ten and seven; and, by the commutative rule, so is 7 plus 10.

The 9's facts follow immediately from the 10's facts. If you know that 10 plus 7 is ten and seven, then (since nine is one less than ten) 9 plus 7 must be one less than ten and seven, or ten and six (16). The 9's facts are more difficult than the others discussed, but, with help, children can see a logical way to remember and figure out the 9's facts.

Two other sets of facts that children generally learn easily are the doubles and the sums to ten. Apparently the doubles are easy because there is only one addend to remember, and the sums to ten are easy if children have learned the finger sets—the number up plus the number still down make ten. Although these two sets of facts are harder to learn and remember than those mentioned above, they usually are learned fairly easily by most children.

All of these sets of facts, with the exception of the doubles facts and perhaps the sums-to-ten facts, provide a logical way for children to understand and remember them. An important part of the reality of human beings is that they have logical minds and can use logic to help understand and remember. Any procedure for remembering important facts that does not take this human trait into consideration is likely to be less effective than procedures that do. So, as well as starting from the learner's physical reality, we should rely heavily on the learner's mental reality as we try to help people understand the abstractions of mathematics.

Unfortunately, a commonly used method of organizing addition facts is by their sum. Children learn the facts that sum to five on one day, to six on another day, and so on. Since there is usually no good logical principle for finding the answer from the question, most children have a very difficult time remembering the addition facts this way. This is a case of ignoring the needs of the learner, but still appearing to have organized things in a logical way.

It is also worth noting that we started with the actual objects to be added, then used other physical objects to represent them. Next, we could use pictures or other seemingly more abstract "objects" and finally we would end with abstract symbols. Tally marks and similar semi-concrete symbols provide another small step that helps children progress from the concrete to the abstract. Such small steps are desirable even though they are not spelled out in detail here.

Addition Table

After they have derived the addition facts from their reality and practiced for some time, children should be encouraged to organize their information. An addition table offers an excellent way to organize the information and to identify difficulties some children may have. Some work is needed to understand the table in the first place, but it is well worth the effort because this is such a common way to organize all sorts of data.

Consider the addition table for sums up to 20 (Figure 3.1). If children are told that they must memorize the 121 facts shown in this table they tend to be a bit unhappy about the situation. However, if they have been shown how to relate facts to things they already know, and have been given a bit of help with some of the other facts before being confronted with the table, they find the task relatively easy.

FIGURE 3.1
Addition Table

0	0	1	2	3	4	5	6	7	8	9	10
0	0	1	2	3	4	5	6	7	8	9	10
1	1	2	3	4	5	6	7	8	9	10	11
2	2	3	4	5	6	7	8	9	10	11	12
3	3	4	5	6	7	8	9	10	11	12	13
4	4	5	6	7	8	9	10	11	12	13	14
5	5	6	7	8	9	10	11	12	13	14	15
6	6	7	8	9	10	11	12	13	14	15	16
7	7	8	9	10	11	12	13	14	15	16	17
8	8	9	10	11	12	13	14	15	16	17	18
9	9	10	11	12	13	14	15	16	17	18	19
10	10	11	12	13	14	15	16	17	18	19	20

Before starting this activity, I usually tell the children the hardest fact for me to remember was always 6 + 8 (or 8 + 6). Then we figure out the answer to this problem, and I repeat the question at odd moments while studying other facts. By the time children have learned all the other facts, they also know that 6 + 8 = 8 + 6 = 14. This trick will work only if a single fact is picked, and it should be reserved for the one fact that children are likely to find most difficult.

Now, examine the table to see if anyone knows all the facts in any one row or column. Children quickly see that they know all the facts in the 0 row and column, in the 1 row and column, and in the 10 row and column. With a bit more effort, and perhaps some practice, they admit to knowing the 2's facts. Then, the 9's facts (because of their relation to the 10's facts). After practicing and checking to be sure everybody knows these facts, we cross them off the table and count the remaining facts. Children are pleased to discover that of the original 121 facts, they already know 85, and have only to work on the 36 facts in the small square of the 3, 4, 5, 6, 7, and 8 rows and columns.

The 3's facts, the 5's facts, and the doubles will require a few days, but children should become proficient with them fairly quickly if the procedures suggested earlier are followed. The near-doubles (one more than double, such as 3 + 4, 7 + 8, 6 + 5, and so on) come next and may require another day or two; but this practice will reinforce the doubles facts, as well as some of the previously learned facts.

That leaves only four facts (and their commutative versions): 4 + 6, 4 + 7, 4 + 8, and 6 + 8. But we've been practicing 6 + 8 and 8 + 6 all along, so there are really only the three 4's facts. We have reduced the chore of learning 121 addition facts to learning three. There are various ways to help children remember those three. If finger sets are used, the fact that 4 and 6 are complements with respect to 10 may help with 4 + 6 and 6 + 4. Then, 4 + 7 and 7 + 4 must be one greater than 10, or ten and one, and, by breaking up the 4 of 8 + 4 into two 2's, we can help them see that 8 + 4 = 8 + 2 + 2 = 10 + 2 = ten and two (or 12).

Thus, by organizing information in a way that is natural to the learner, and helping the children see connections, we can greatly simplify an apparently overwhelming task. There are, of course, other specific ways to achieve essentially this same result.

Practice: A Game

No matter how well children understand the basis for the addition facts, they will need lots of practice to make them automatic. Unless these facts are automatic, children will use their time and creativity to reconstruct the addition facts when they should have gone on to bigger and better things. Thus, practice is essential. As suggested earlier, practice need not be of the "drill and kill" variety provided with flashcards, ditto sheets, and other dreary training techniques. Instead, various games and other activities can be used. Beyond that, to help the teacher assess the status of individuals with respect to each fact, as well as to provide further practice, some full-class response activity is appropriate. In small doses, with positive reinforcement from the teacher, such activity is usually helpful.

Many games can be used to practice addition facts. Let's consider a simple example. Provide each child with two cubes. On one cube will appear the numbers 0, 1, 2, 3, 4, and 5. On the other cube will appear the numbers 5, 6, 7, 8, 9, and 10 (5 appears on both cubes). Later we will show how these same cubes can be used in a full-class response drill.

Have the children play Cube–15 in pairs (they may play in triples if this suits the situation better, and you may even want one member of a triple to act as referee if the children are not sufficiently mature to play the game in an appropriate manner in pairs). Each pair of children will have four cubes—two 0–5 cubes and two 5–10 cubes.

The first player rolls any cube, then rolls a second cube and announces the sum of the two numbers rolled (if the first two numbers rolled are 3 and 4, the sum so far is 7). Then, the player may stop or roll a third cube, adding its number to the previous sum (if the next number rolled is 9, the sum is 16). Again, the player may stop or roll the fourth cube, but must stop after the fourth cube. The goal of the game is to get a final sum as close to 15 as possible.

Think about the game. Could the first player have rolled a 2 on the third roll? Why not? If you were the first player, would you stop after the third roll (with a score of 16)? Why? How close is 16 to the goal? Remember, it is all right to go over 15, so 16 is a good score since it is only 1 away from the goal of 15, and would tie a score of 14 (and beat any score other than 14, 15, and 16). If you were playing the game, would you roll the cubes in the same order the first player did? Why or why not? If you had a score of 13 after rolling the two 5–10 cubes (and didn't know what the other player's final score would be), would you roll one of the remaining cubes? Why or why not?

Notice that even though the ostensible purpose of this game is to practice addition facts, there is a great deal more involved. To decide who won, the players must do some informal subtraction. To decide whether to roll a cube in a given situation, or which cube to roll in cases where there is a choice, some informal probability may be used.

Beyond that, children generally enjoy playing games of this sort, and therefore continue practicing long after they are no longer required to do so. If practice is essential, we should try to provide practice that the learners will enjoy so they will continue practicing. Some children do enjoy practicing with flashcards or electronic equivalents of flashcards, and there is no harm in letting them practice with such devices if they enjoy it. But a well-devised game can often provide a great deal more than just enjoyable practice. In this case we have seen that it integrated subtraction and probability into the addition practice.

Perhaps the most important experience provided by the Cube–15 game, and other games like it, is the opportunity for students to recognize and formulate their own problems. Throughout formal school mathematics, from kindergarten to graduate school, students are asked to solve somebody else's problem or prove somebody else's theorem. They almost never formulate and solve their own problems. Therefore, mathematics classes are less interesting than they could otherwise be, and graduates of such programs are ill equipped to use their mathematics in the real world.

In the real world, we find ourselves in situations that seem to call for some mathematical analysis (for example, what time do I have to leave home in order to get to work on time?). We may or may not formulate the problem explicitly. Then we bring to bear whatever skills and knowledge we have to help solve the problem. The recognition and formulation of the problem are the important steps. Then we conjecture and test possible solutions, and in many cases try to convince somebody else that our solution is correct. In Cube–15 and similar games, the teacher does not tell the players that they must find a better way to play the game, nor even ask them questions like the ones I asked here. The players want to win and therefore are inclined to use mathematical thinking to improve their chances of winning.

Beyond that, after discovering what they think is a better strategy, most players take a certain amount of pride in telling others about the new, good strategy. This enhances communication skills, and it encourages independent thinking on the part of both participants. If the teacher tells a pupil what the best strategy is, the pupil is inclined to accept the strategy without thought because the teacher is the authority. If, on the other hand, another pupil suggests such a strategy, the pupil has an inclination to think about it and challenge it. Thus, both discussants remain free to think critically rather than accept the proclamation of an authority.

Children ought to play games of this sort with different partners so as to encourage the interchange of different ideas. Teachers should watch children play such games as much as possible, both to see what kind of thinking is going on and to be sure children are getting the correct practice. Occasionally, both players will make the same arithmetic error (for example, $7 + 8 = 16$) and thus reinforce each other's error rather than correcting it. Sometimes a child, while concentrating on the game, will make an error that has not been obvious to the teacher at other times. Observation of game-playing activity resembles observation of real-life-out-of-school activities as closely as anything we are likely to see in school. Such observation will often give greater insight into a child's thought patterns than anything else the teacher can do.

Practice: Response Drills

A second kind of practice is the whole-class response drill. Teachers have used such drills, in one form or another, for hundreds of years. One of the most common forms of whole-class response drill is the oral drill. The teacher says or writes a problem and the class answers orally in unison. The most obvious difficulty with this form of response drill is that the teacher cannot tell for sure what various children are saying, and, in fact, can't even tell for sure whether individual children are participating. A child in the back of the room may be saying something quite different from what the rest are saying without the teacher's realizing it, or the child may simply remain silent (perhaps moving his or her mouth).

Another form of whole-class response drill that was popular during the 19th century requires each child to have a small slate and a piece of chalk. The children write their answers on their slates and then show the teacher their responses. Such slates are easy to find in antique stores, and some educational vendors and hobby stores again have them available for schools. Writing answers on a pad of paper, slate, or similar device is possible, but there are disadvantages even beyond the need for equipment. Because children can easily see each other's responses if they are written large enough for the teacher to see, the independence of a pupil's work is hard to judge. Beyond that, children's writing often is not sufficiently legible for the teacher to be sure what answer was meant.

Various electronic devices also are available. With such a device, each child has a small keypad, and the teacher can monitor each child's response. The teacher can get a percentage score of the entire class for certain answers and can obtain other useful information. These devices seem to be quite effective for some purposes, but they have been expensive and are not commonly used.

Number cubes, like those used in the Cube–15 game, also provide an easy way to show answers. If the cubes are at least two centimeters on a side and the numerals are written as large and legibly as possible, a teacher can easily tell, from a distance of 15–20 feet, what answer each student is showing—or even whether all children are participating. For numbers greater than 10, of course, each child needs more cubes. Two additional cubes with the numbers from 0 to 10, but with a small "TENS" written below the numeral allow children to show numbers up to 100 and encourage further thinking about the base ten nature of our numeration system.

For numbers greater than 99, a stiff cardboard or plastic card with number wheels attached by grommets so they can turn and show only one digit at a time can be very effective. Four or five such number wheels per card can be used to show numbers as great as 9,999 or 99,999. They can also be used to show decimals (have children point with their fingers where the decimal point should be), fractions (agree that the first two digits will show the numerator and the last two the denominator, so that 0508 shows 5/8), time (0935 stands for 9:35), and other answers.

For some kinds of responses, one of the most effective and convenient response devices is the hand. For example, if the teacher asks a "yes" or "no" type question, holding the hand clenched with the thumb pointed up could mean "yes" and the thumb pointed down could mean "no." If a third response is needed (such as "There is not enough information to answer," or "I don't know"), the open hand held horizontally can indicate that. Similarly, for estimation problems, thumbs up can mean the answer is greater than a certain predetermined number, thumbs down can mean it's less than the number, and a horizontal open hand can mean it's too close to call. Hand signals should always be sufficiently different from each other so the teacher can distinguish one from another easily. Thus, using fingers to show numbers from 0 through 10 is possible, but unless responders hold their hands steady with their fingers well spread apart, the teacher may have a difficult time deciding whether the right answer has been shown.

Whole-class response activities encourage practice, allow students to correct their own errors, and allow the teacher to identify difficulties that individual students are having or that are common to the entire class. To allow students to correct their own errors, the teacher should respond with the same device the children use (slate, paper, cubes, cards, thumbs, or whatever) and also, at the same time, give the correct response orally, so that children's answers are reinforced or corrected both visually and aurally.

Applications to Interesting, Real Situations

Once children have explored a concept through their own experiences, abstracted it, and practiced it until they have developed an appropriate level of skill, they should apply it in a variety of settings. Where possible, applications should raise questions for which children might want to know answers.

Interesting games or puzzles are perfectly good applications, even if they may not seem really important in the adult world. Although "word" problems provide a reasonably efficient method of having children do lots of "realistic" problems in a hurry, they often are not at all real to children. They have the added disadvantage that they often are presented in such a way as to make children solve them without really thinking (e.g., using the most recent algorithm taught, using the same algorithm as the other problems on the page, looking for "key" words, and guessing from the size of the numbers what operation to apply).

Because of these inadequacies of word problems, children should regularly be expected to use their skills to solve problems that are not neatly written in the textbook or on a ditto sheet. Such problems might include projects (What's the shortest route home? How many heel-to-toe steps will Maria take to walk the distance of three tables put together? How much will it cost to buy three candy bars? How many hours a week do I watch television? and so on). They might involve matters of importance largely to adults (How can we "balance" a checkbook?). But children should, in general, be exposed to applications of their mathematics that came from somewhere other than the textbook or a ditto sheet.

Example 2: Multidigit Addition

The story told earlier about Max Beberman and his daughter, who was learning two-digit subtraction, demonstrates the kind of trouble children have with multidigit algorithms when insufficient attention is given to the learner's physical and intellectual status. An early emphasis on the use of symbols, as opposed to development of understanding based on concrete situations, is inappropriate and often results in misconceptions, rapid forgetting, an inability to apply mathematics to real situations, and loss of desire to learn or do mathematics.

Two-Digit Addition, without "Regrouping"

One of the most common mistakes with two-digit addition is to try to teach the "easy" part early and save the harder part for later. So, in 1st grade, children are often taught to add two-digit numbers "without regrouping" or "without carrying." This makes it appear that children have learned to do two-digit addition; and parents, children, teachers, testmakers, textbook authors, and others are happy that the children have progressed so far. What is really happening is that children are getting a gross misconception of the process.

If a child learns to add two two-digit numbers such as 35 and 54 (where the sum of the units digits is less than 10) in the usual manner, the child has every reason to suppose

$$\begin{array}{r} 35 \\ +54 \\ \hline 89 \end{array}$$

that the procedure simply consists of adding the 3 and 5 to get 8, and also adding 5 and 4 to get 9—two separate single-digit problems. It is no wonder that the same child, faced with adding 35 and 57, gets 812 and assumes this is reasonable since the addition done in school has nothing to do with reality anyway. They add 3 and 5 to get the 8, and then add 5 and 7 to get 12.

$$\begin{array}{r} 35 \\ +57 \\ \hline 812 \end{array}$$

Even when teachers are convinced that adding two-digit numbers without regrouping is undesirable in grade 1, they will tell you they must teach it because it's on the year-end standardized test. Some years ago I was trying some materials that avoided two-digit addition until grade 2. I told the teachers the children would do well enough on everything else to make up for their inability to do two-digit addition. But at the end of the year, the experimental groups did as well on two-digit addition as control groups.

I suspected the teachers of teaching two-digit addition despite my pleas. But I was teaching one of the classes myself, and my class did as well as anyone else's. So I asked some of the children how they did a problem like 35 + 54 (the problems on the test were arranged in vertical format). The first child examined the problem for a moment and wrote down the answer. I asked "How did you do that?" He looked at the problem more carefully and said: "I added 3 and 5 to get 8. Somebody forgot to put in the other addition sign, but I knew I was supposed to add 5 and 4 to get 9."

The moral of this is that children don't have to be taught to do such problems the wrong way—they can figure that out all by themselves. But if they are encouraged to practice a procedure that is incorrect, they will become quite good at it, and will be loath to give it up. Telling them things like "Start on the right" or "Remember to add units to units and then tens to tens" doesn't do any good, because children will be happy as long as they are using a procedure that is consistently marked correct, even if it is based on a misconception. By the time they meet problems where the sum of the units digits is greater than 9, it is too late.

Certainly, there is no practical reason for teaching 1st grade children addition (or subtraction) of two-digit numbers without regrouping since they are just as likely, in real life, to meet those with regrouping as those without. If such teaching has the disadvantage that children practice doing something they shouldn't be doing (adding and subtracting as though the units digits and the tens digits of numbers have nothing to do with each other in these operations), then we ought to forgo the activity until such time as it can be done correctly.

In the meantime, if teachers want children to do some adding and subtracting involving two-digit numbers that will be beneficial, they should do problems like 18 + 2 by having the children "count on." In this case, the child starts counting with 19, then 20, and stops: 18 + 2 = 20. When the small number is no larger than 3, most 1st grade children have no trouble doing such problems. Using a number line or base ten materials may be helpful. If appropriate examples are used, children quickly discover that sometimes what is added in the units column affects the tens column and sometimes it doesn't. This is exactly what they should learn. Beyond that, they practice a procedure (however inefficient) that will work to solve any addition problem. Counting back for subtraction is equally effective.

Two-Digit Addition, in General

Sometime during 2nd grade, most children are ready to learn formal two-digit addition and subtraction procedures. Children should know the addition and subtraction facts well; they should be very comfortable with numbers at least up to 100 and preferably up to 200; they should have had a lot of experience with base ten, both with concrete objects and with base ten symbolism; and they should be sufficiently mature so that they can see connections and keep a reasonably complicated procedure in mind as they carry it out.

Then, start with a physical problem in which things are naturally grouped by tens. For example, three children are holding up all 10 of their fingers, and another child is holding up 7 fingers (37). In another group, two children are holding up all 10 of their fingers, and a third child is holding up 6 fingers (26). How many fingers are up altogether? (37 + 26 = ?)

The children can see 5 tens. Bring the other two children together. They have 13 fingers up. If you pretend to remove 3 fingers from the six-child and place them on the seven-child, then the first now has 3 fingers up and the second has 10 fingers up. Now there are 6 tens and 3, or 63. A similar procedure can be used with ice cream sticks in much the way Max Beberman did subtraction with his daughter.

As with addition facts, children should be encouraged to use physical representations of objects when the objects themselves are not convenient to manipulate—use ice cream sticks or fingers to represent cars, houses, or people, for example. Then, as another step toward abstraction, they may use pictures or other symbols.

After children have done many examples of the sort described, using several different sets of physical objects, as well as some physical and symbolic representations of physical objects, they should be encouraged to keep written records of what they are doing. For the finger problem mentioned previously (37 + 26), for example, encourage the children to start by lining up the

$$\begin{array}{r} 37 \\ +26 \\ \hline \end{array}$$

tens and ones in a vertical format because that makes it easier to keep track of the activity (though it is not essential, and if children prefer not to do it this way, encourage them to do it in whatever ways seem appropriate).

What did we do first? Added 3 tens and 2 tens getting 5 tens. Let's write that in the tens column. Next, we added 7 and 6, getting 13. We rearranged the 13 into 1 ten and 3. That made 6 tens and 3 or 63. Notice that the natural way to keep records here is to start on the left and work to the right. That is entirely acceptable but requires a bit more work than if we started on the right and checked first to see if adding the units would produce an extra ten. Since 7 + 6 yields 1 ten and 3, we will have an extra ten. Now, we can simply write the 3 in the units column, add 1, 3, and 2 tens, and write the 6 in the tens column.

$$\begin{array}{r} 37 \\ +26 \\ \hline 5 \\ 13 \\ \hline 63 \end{array}$$

Slightly less work and mess is required if we start at the right. This is enough reason to convince most children that they ought to start at the right rather than the left. If they choose to work from left to right they should be allowed to do so as long as they can do it correctly. There's nothing inherently wrong with working from left to right—it just requires more work and produces a slightly messier paper.

Notice that there is a reason for doing things the way we usually do them. In the typical classroom the pupil is told to work from right to left when adding multidigit numbers because the teacher or textbook said to work from right to left. Then, when subtraction is introduced, they work from right to left because that's the way we did it in addition. In multiplication we work from right to left because that's the way we did it in both addition and subtraction. In division, we forget all this because there we work from left to right.

No wonder children begin to get the idea that mathematics is a form of mysticism in which they must carefully follow rules handed down from generation to generation to get answers that will satisfy the generation that precedes them. No wonder children grow up thinking that there is a mathematics done in school, and there are real problems done outside of school, and the two have nothing to do with each other. There are reasons for almost everything we do in mathematics. Children should be encouraged to discover, or at least see, those reasons. Understanding the reasons will help them remember how to do the mathematics; but more important, it will help them understand that the mathematics is related to their reality and that mathematics can be used to help them understand the real world.

We should now practice two-digit addition problems with response exercises, with games, with real applications that are appealing to children, and so on. One of many games that can be played to practice two-digit addition is "Roll-a-Problem: Two-digit Addition." In this game, the teacher, or one member of the class, rolls a cube (say the 0–5 cube) four times and each player writes the number in one of the four spaces. The cube is rolled again, and again each player writes the number in one of the remaining three blanks, and so on, until all four blanks are filled. Then, each player adds the two two-digit numbers formed, and the greatest total wins. In a class of 30, of course, there usually will be lots of people with the best possible score. In smaller groups, ties are less likely.

```
    _ _
+   _ _
    _ _
    _ _
    ___
```

Notice that, as in Cube–15, there is more to the game than just practicing addition of two-digit numbers. Deciding who won requires comparing numbers (a not altogether trivial activity for a seven-year-old), and to avoid a certain amount of embarrassment, children have a strong tendency to check their answers before announcing the results—a generally felicitous development.

But beyond that, there is a good deal of probability and strategy inherent in this game. Suppose the 0–5 cube is being rolled and the first number rolled is a 3. Where would you put it? Does it make a difference whether you put it in the top or the bottom tens spot (or units spot if you chose to put it in a units spot)? Would the number of people playing change your decision? Would your assessment of the skill of the other players change your decision? The thinking involved can become quite sophisticated even for an apparently simple game like this.

One more point should be made about teaching both single-digit and multidigit addition. Whenever possible, addition and subtraction should be taught at about the same time so that learners can contrast the two and so that problems can be presented that are not all solved by using the same operation.

Often, when addition and subtraction are taught at about the same time, teachers complain that children are confused. It is true that children find it much easier to learn only one algorithm at a time, then practice that algorithm for a long time, then solve "problems" using only that algorithm, and then go on to something else. The difficulty with such a procedure is that it pretty much eliminates the need for the learner to think. When the time comes to decide which operation is appropriate, the pupil is unlikely to have given enough thought to what the operations mean to be able to make intelligent decisions. Thus, by saving "confusion" and work at the beginning, we do the learner a disservice that doesn't become apparent until later. This matter is discussed further in Chapter 4, "Problem Solving."

Three-Digit (or More) Addition

If children really understand the base ten numeration system and the process for adding two two-digit numbers, three-digit addition should seem quite natural to them. Using base ten blocks, ice cream sticks with bunches of ten and bunches of ten tens, and other similar physical materials, they should be able to follow through essentially the same steps used for two-digit addition, including the abstracting involved in keeping records, practice through various means, and good applications.

After that, learning how to add in general should be almost automatic if children have understood what went on before and if the general algorithm is approached appropriately. If you are in the middle of an addition problem such as the one shown here, you don't have to know whether the 8 and 5 are units, tens, hundreds, or something else. You can think of them as units (except for possible "carrying" from the column to the right). If they are really hundreds, there would be 8 hundreds and 5 hundreds to be added, giving 13 hundreds (or possibly 14 hundreds if there were more than 9 tens). Write the "3" in that column and add the "1" in the column to its left getting 6 ten-hundreds (thousands) (or 6 ten-whatever-is-in-the-right-most-column-shown). In this case, we can think of whatever is in the right-most column as units, bunch ten of them together to get the things represented in the second column, bunch ten of those together to get the things represented in the next column, and so on. Doing this physically may be desirable for some children, but *imagining* this activity is more appropriate where possible.

$$\begin{array}{r} \ldots 738 \ldots \\ + \ldots 425 \ldots \\ \hline \end{array}$$

By doing enough examples of this sort and encouraging children to see that the procedure is the same no matter which column is involved, you can quickly show them how to add, no matter how big the numbers. Of course, a great deal more development is required than is shown here, but there is no reason why a typical 3rd grade child cannot understand and use a general addition algorithm if the developmental work is done properly.

The generalization is important for several reasons. First, it saves a lot of work. In many classes and many textbooks, children are taught the algorithm for two-digit numbers, then three-digit numbers, then four-digit numbers, and so on, over a period of years as though each procedure were different from the others. This is a waste of time and gives a distorted notion of mathematics.

Second, if approached correctly, a general addition algorithm shows the beauty and power of base ten notation. The fact that a digit in any column stands for a number exactly ten times what the same digit in the column to the right would stand for is the crucial point here. Since that is always the case, you don't really have to know in which column you're working to proceed.

Third, this shows the great power of abstract reasoning and generalization. By thinking abstractly and in general terms, we can make a very powerful statement that allows us to do many more problems than we would be able to do without such abstraction and generalization.

Another important point to be mentioned here is that this discussion of addition extends over a period of four to five years in the life of a child.

What happens in kindergarten and grade 1 very much influences what can be done in grades 2 and 3. If each step is taken in a way that reinforces children's outside knowledge while gently correcting misconceptions, there will be regular constructive growth. Each step should be planned with a knowledge of what has gone on before and what will come later.

One of the great weaknesses of U.S. education in mathematics is that teachers seldom know what happened in previous grades or what will happen in subsequent grades. A good textbook series should correct this by relating the development in one grade to previous and subsequent grades and by telling the teacher (in the Teachers' Guide) how a particular activity fits with previous and subsequent work. A supervisor should also know the entire curriculum well enough to understand the "vertical" connections and to encourage teachers to teach in a way that will facilitate subsequent teaching (see Chapter 6 of this book, "Connections").

The discussion here has been limited mostly to computation. Everything discussed is applicable, with appropriate modifications, to other topics in school mathematics. There are several reasons for using computation as the prime example:

1. The content is familiar to everybody, and therefore easier to follow than some other topics might seem.

2. Because we have been teaching computation in relatively less effective ways for decades, showing the details of better ways to teach computation should allow the reader to see the contrast between the more and less effective procedures.

3. Since the reformers who are advocating introduction of new topics (e.g., data analysis, estimation, discrete mathematics, earlier study of geometry) are also advocating a better pedagogy, textbooks and other instructional materials from which these topics might be taught are more likely to approach them in the desired manner.

4. One such topic (functions) is considered in Chapter 6 to show that both traditional and nontraditional topics can be treated this way.

References and Further Resources

Conference Board of the Mathematical Sciences. (1983). "The Mathematical Sciences Curriculum K–12: What Is Still Fundamental and What Is Not." In *Educating Americans for the 21st Century. Vol. 2. Source Materials*, edited by the National Science Board Commission on Precollege Education in Mathematics, Science and Technology. Washington, D.C.: National Science Foundation.

Dossey, J.A. (1989). "Transforming Mathematics Education." *Educational Leadership* 47: 22–24.

Driscoll, M. (January 1988). "Transforming the 'Underachieving' Math Curriculum." *ASCD Curriculum Update.*

Mathematical Sciences Education Board. (1989). *Everybody Counts.* Washington, D.C.: National Academy Press.

National Council of Teachers of Mathematics. *(1989). Curriculum and Evaluation Standards for School Mathematics.* Reston, Va.: NCTM.

National Science Board Commission of Precollege Education in Mathematics, Science and Technology. (1983). *Educating Americans for the 21st Century*. 2 vols. Washington, D.C.: National Science Foundation.

Willoughby, S.S. (1981). *Teaching Mathematics: What is Basic?* (Occasional Paper #31). Washington, D.C.: Council for Basic Education.

Willoughby, S.S., C. Bereiter, P. Hilton, and J.H. Rubinstein. (1981, 1985, 1987). *Real Math, K–8 Textbook Series*. La Salle, Ill.: Open Court.

4

Problem Solving

The latest fad in education is "critical thinking and problem solving." Courses in "How to Think—About Nothing" have been propagated. Books, pamphlets, and courses purporting to teach anybody how to solve any problem are abundant.

Recently a teacher approached me at a mathematics convention and asked what I would recommend to teach problem solving. I suggested a teacher. She explained that she wanted a pamphlet or book she could use on Friday afternoons to teach problem solving. I wondered aloud whether, if mathematics (and other subjects) were taught properly in the first place, such a book, and such Friday afternoon sessions, wouldn't be superfluous. She chose to speak to somebody else. I'm sure she found what she was looking for.

There is no royal road to critical thinking. There's not even a pauper's paved path to easy problem solving. Teaching today's children to become the thinking, caring leaders who will be able to solve the world's increasingly complex and quantitative problems requires a total commitment, not just a Friday afternoon contribution.

Possible Pitfalls in Teaching Problem Solving

Because there are so many people willing to provide easy nostrums to teach critical thinking and problem solving, there is a need to explain what's wrong with some of these methods before describing some of the comparatively more difficult, but more successful, procedures that help students become better critical thinkers and problem solvers.

Rules

Teachers often have a misguided belief that we benefit pupils by doing their thinking for them. If we can come up with a set of rules that children can memorize to solve mathematics problems, to do science, to write coherent essays, and to understand the broad implications of history, then our pupils will be spared the unfortunate necessity of actually thinking for themselves. In fact, if such rules actually did exist, it would be possible to program a computer to do these higher-level thinking tasks. We must remember that educating a human being is a very different task from programming a computer, and a much more difficult task.

A 3rd grade textbook presents the following four rules for solving problems: (1) Read the problem. (2) Think. (3) Add or subtract. (4) Check your answer.

Later in the same book, the following *three* rules for solving problems appear: (1) Read the problem. (2) Add, subtract, or multiply. (3) Check your answer.

The astute reader will notice that if there are three operations from which to choose, the authors believe you no longer should have the luxury of thinking.

These may seem to be unusually naive sets of rules for solving problems, but they have flaws that are common to most. The rules are essentially not helpful, and are probably counterproductive. If the problem does not come in the form of a written "word" problem, of course, reading will not necessarily be an appropriate activity. But even a more sophisticated instruction, such as "Be sure you understand the situation" is not likely to be of much help without a great deal of explanation and a lot of experience. Such instructions simply don't mean anything to children. If children are required to memorize such rules, they will do so to regurgitate them on a test, but experience shows that they never actually use such rules to help them solve real problems. This is not surprising, because the rules obviously wouldn't be of much help.

The injunction to "think" is of no value, of course, without some clue as to what the topic of thought might be, or how one might go about thinking in a constructive manner. That probably accounts for the fact that nobody appears to have noticed the absence of this precept when the rules were repeated.

The next rule, "Add or subtract (or multiply)," shows clearly the degenerate nature of such rules. There is an inherent assumption here that all problems involving number are to be solved by performing some mathematical operation and that the appropriate mathematical operation will always come from among those the pupil has been taught—usually quite recently. That simply is not the way the real world works. After teaching multiplication to a group of students for a few weeks, try giving them the following problem:

If 1 man can jump a stream that is 3 meters wide, how wide a stream can 5 men jump?

Choosing the most obvious arithmetic operation is not necessarily the best way to stay dry in this case.

The last rule, "Check your answer," is a grossly oversimplified version of a useful custom. As is discussed later, most good problem solvers make it a habit to think about a problem and their solution to the problem after they have solved it. This includes checking to see that the answer seems reasonable, but involves far more. It includes trying to find a better way to solve the problem and trying to generalize the problem, the answer, and the method of solution. It includes trying to think of other problems that might be related, and so on. "Check your answer" simply does not convey the richness of this activity when carried out properly.

Key Words

As problem-solving mania spread through the educational world, publishers produced a plethora of books and pamphlets on how to teach problem solving (on Friday afternoons after the real work of the week was completed). An advertisement for one such booklet presented the following problem and solution procedure as an example of the wonderful things the booklet could do for education:

Jackie's dress size is 8. Her sister's dress size is 14. How many sizes smaller is Jackie's dress than her sister's?
Step 1: Circle the key word.
Step 2: Circle the correct equation:
$14 + 8 = 22$
$14 - 8 = 6$
Step 3: Write your answer.

Key words are used for the purpose of avoiding thought. Children can start at the end of a problem, work their way backwards until they reach a "key word" and then do the appropriate operation on the two numbers given (if there are more than two numbers the operation is almost certainly addition, so no key words are needed). In this case, the key word is "smaller" and having circled that word should lead the child to also circle the second equation (involving subtraction) and arrive at the answer, 6.

What's wrong with this procedure? Almost everything. Try the following problem in a 2nd or 3rd grade mathematics class:

Mary walked 11 meters north. She then turned and walked 7 meters west. Did she turn right or left?

If the most common answer is 4, you will know that the class members have mastered the key word procedure (since "left" always means subtract) but are not reading and thinking about the problem.

Step 2 of the dress problem is based on the same assumption we discussed in connection with rules, namely that all mathematical problems are solved by applying an arithmetic algorithm that the problem solver has been taught. In this case, apparently the only operations available are addition and subtraction, so there are only two possible equations (presumably $8 - 14 = -6$ is not to be considered for other reasons).

Finally, this paragon of problem-solving pamphlets has led the innocent reader unerringly to a wrong answer. The correct answer to the problem, as anyone who has even a passing acquaintance with women's dress sizes knows, is 3, not 6. Women's dress sizes come in even numbers

(Odd-numbered dress sizes are called "juniors" and are not usually worn by people who use the even-numbered sizes.)

The point of this comment is that dress sizes are not numbers in the usual sense. Numbers can be added, subtracted, multiplied, and divided; and the result will make some sense. Highway names are a good example of symbols masquerading as numbers that aren't really numbers. As a car approaches Hartford, Connecticut, the motorist sees a sign that says "86 is now 84." Every time I see the sign I absently wonder whether that means that 86 is now divisible by 7.

In the same sense that we do not do arithmetic with highway numbers, we do not do arithmetic with dress sizes. If you were to cut up Jackie's sister's dress (size 14) to make a dress for Jackie (size 8) you would not have three-fourths of a dress for Jackie left over.

Dress sizes are not even universally ordered in the natural way. If you are size 14 and would like to be size 12, you can accomplish that noble goal either by dieting, or by buying your dresses in a more expensive store. Alternatively, if you can wait about 25 years without putting on weight, you should accomplish the same feat since the size assigned to a given dress appears to be reduced about one step every 25 years. So, in the 21st century, the equation $8 - 14 = -6$ may no longer be an unreasonable way to determine a dress size.

The potentially exciting information that could be communicated to children with the dress problem is that there are lots of things around that seem to be numbers that do not behave at all like numbers. We must be careful how we treat those things. If a child cannot reach the "20" button on a hotel elevator, pushing the "10" button twice will not make the elevator go to the 20th floor, nor is it likely that a person who is staying in room 1999 of that hotel is next door to the person in room 2000.

To carry out the human activity of solving problems, we must think about situations and use reasonable judgment.

Unrealistic Problems

During my second year of teaching, a boy named Bill was in one of my 8th grade mathematics classes. Bill was very good at arithmetic but could not solve word problems. He would say "I can do the mathematics, Mr. Willoughby, I just can't do the problems." I tried many times to explain that there was no point in learning "the mathematics" (or symbol pushing) if you couldn't use it to solve problems.

Bill was on the volleyball team that I coached. One evening at about 6 p.m., Bill and I were the last two people to leave the gymnasium, and therefore the last two people to leave the school. As I heard the door close and lock behind us, I realized that it was cold and snowy out, that I had left my coat in my locked classroom at the other end of the building, and that I had no way of getting back into the school. I stood there and shivered for a moment.

Bill said, "Is something wrong, Mr. Willoughby?"

I explained the problem to him and said it would be a long cold walk down to the bus stop. He asked, "Would you like your coat, Mr. Willoughby?"

I said, "Yes."

Bill disappeared around the corner of the building, apparently not hearing when I yelled, "Bill, don't you want the key to my classroom—it's locked."

About three minutes later Bill walked through the gym door with my coat. He handed it to me. I said, "Thank you."

He said, "You're welcome." We never discussed the matter again.

As I walked down to the bus stop in my nice warm coat, I thought about "poor dumb Bill" who couldn't solve problems. I began to wonder whether it was Bill who was out of step or the schooling with which he was trying to cope. After reading and thinking about some of the "word problems" that appeared in our textbook, I concluded that anybody with even a modicum of intelligence and good judgment would not waste time trying to deal with such stuff. The following example comes from a fairly recent first-year algebra book, but is quite similar to the problems with which Bill was asked to deal.

> Mary's mother needs three hours to do the laundry. If Mary helps her, they can do the laundry in only two hours. How long would it take Mary to do the laundry by herself?

This problem was obviously written by someone who had never done the laundry. Doing the laundry is usually not a two-person activity. If it takes Mary's mother three hours to do the laundry, it will probably take at least that long for Mary and her mother to do the laundry together. Sorting the laundry requires one mind in control or the sorting criteria will change from time to time and make for a very inefficient process. Once the laundry is sorted it is placed in a machine with various quantities of soap and other chemicals. Then, the person or people wait while the machine goes through its cycles. Having two people wait does not make the agitator go faster than if only one person is waiting. The damp, soggy mass is then transferred to a dryer where it rotates again, with no noticeable difference in speed dependent on the number of people waiting. The clean, dry laundry is again sorted. The launderers then search for the missing sock—*that* may be a two-person activity. And how long it takes Mary herself to do the laundry depends of a lot of things—her age, her knowledge of laundry intricacies, and how much allowance she gets for the chore.

The point is that most of the problems presented to students in mathematics classes are patently unreal. Rather than motivating the students to solve problems and study more mathematics, these problems teach what the brighter students already suspect, namely that mathematics has nothing to do with the real world and can be safely ignored by anybody who wants to have a better understanding of the world.

Nonpertinent Clues

For his Ph.D. dissertation, Stanley Erlwanger (1974) interviewed children about how they solve problems. One boy who had done quite well at solving problems throughout his school career said that if there were two numbers and they were both big he subtracted. If there was one large and one small number he divided, and if it didn't "come out even" he multiplied.

Interestingly, P.R. Stevenson wrote an article for the *Journal of Educational Research* in 1925, in which he reported an almost identical interview with a young pupil of that era.

Some years ago, I told this story to my secondary school mathematics methods class. About two weeks later, one of the class members who was doing her student teaching said, "Do you remember that method of solving problems you taught us two weeks ago? It really works! My students just love it, and my cooperating teacher is going to use it from now on."

Since then, whenever I tell a story that is meant to be funny, I make a point of laughing.

Unfortunately, we as educators all too often lose sight of our real goals. We settle for short-term, intermediate goals such as getting all the kids to do well on a standardized test on problem solving. Scores on such tests presumably are positively correlated with a person's ability to solve problems in the real world. But the more effort we put into teaching for the tests, the more likely it is that we will fool ourselves, the children, and others who are interested, into believing that we have taught the children how to solve real problems when we have not. Some textbooks, and even some tests, seem to have been written by people who have joined in a conspiracy to make it appear that children have learned to solve problems when they have in fact only learned how to take certain kinds of tests. Such activity is harmful largely because it takes time and effort away from the serious goal of helping children understand and deal with the real world. In real life, they will not often meet problems that they can solve only because of the collusion of the problem poser.

Teaching for Problem Solving

The state of California has recently instituted a 12th grade test of mathematical achievement consisting of 12 parts. One of those parts is an essay. On one of the other parts calculators are not allowed. According to a knowledgeable and reliable source, some 12th grade teachers, upon seeing samples of the new test, asked, "How can we prepare students to take this test in just two weeks?" The response was, "You can't. Twelve years of good mathematics education is needed to properly prepare people to take this test."

That's as it should be.

The way in which people originally learn mathematics plays an important role in determining whether they are able and willing to use it subsequently to help understand the world around them and solve problems using their mathematics. Deriving mathematics from the real world, understanding the power of abstraction, practicing, and applying the mathematics to situations of interest will all help people use their mathematics to solve problems forever after. Avoiding some of the common pitfalls described in this chapter will also encourage people to use their mathematics effectively. What more can be done?

Undoubtedly, the most effective way for people to learn to solve problems is for them to solve problems—lots of problems. Beyond that, however, there is now considerable research evidence that people who think about their problem solving *after* they have solved a problem are better problem solvers than those who don't. Although learning rules in advance that are designed to reduce the amount of work needed to solve problems seems to be of little help (and can even be counterproductive), working out rules as one solves problems, or in retrospect, is helpful in solving future problems.

In their early schooling, people should be taught their mathematics in a way that will encourage future problem solving. They should be exposed to many problems. As time passes, they should be encouraged to think about their problem-solving strategies retrospectively. Later, they should be encouraged, with help, to think about general strategies that have helped them solve problems. They should then organize those strategies in a more or less informal way to help them recall appropriate strategies when they meet various types of problems in the future. Of course, they should continue to learn new mathematics in the same constructive manner as earlier and should continue solving lots of problems, using and modifying their strategies as they work.

Practice in Solving Problems

Before we can set out to practice solving problems, we must first learn how to recognize a problem. This is not an altogether trivial task. Let's consider an example.

A hunter started at camp and walked one mile south. Then she walked one mile west. There she shot a bear. The bear was heavy, so instead of retracing her steps, she walked one mile straight back to camp, dragging the bear behind her. What color was the bear?

Is this a problem? Yes and no. If you've seen it before, chances are you remember the answer and therefore this is not a problem at all, but simply a memory exercise. If you've not seen it before, you may be so totally unable to deal with the situation that you have no idea where to start; so you puzzle over the question for a moment or two, conclude that the person asking the question has lost his mind (or perhaps never was totally right in the first place), and you forget the whole thing. Or, if you've never seen the story before but can use your knowledge of the way human beings identify places and directions on the face of the Earth and relate that to colors of bears, this may be a problem for you, and you may or may not be able to solve it.

If you solve the bear problem—or if you've seen the solution previously—you will, I'm sure, be unable to resist trying to generalize the problem to identify all places on Earth where a person could start, walk one mile south, one mile west, then one mile back to the starting spot (there are infinitely many such spots, which are described at the end of this chapter for those who are interested).

The situation in isolation is not a problem, nor is it a nonproblem. To decide whether it is a problem, we need a person to go with it.

Many scholars have done significant research on what constitutes a problem and how to define a problem. For our purposes, we will consider a problem to be a situation in which a person wants to reach a particular goal, is somehow blocked from reaching that goal, but has the necessary motivation, knowledge, and other resources to make a serious effort (not necessarily successful) at reaching the goal.

This definition immediately eliminates most of the activities called "problems" in school mathematics. In many cases the pupil is not the least interested in reaching the goal set by the teacher or textbook. But, even if the goal is of no inherent interest to the pupil, external pressures may produce interest where there was none, so we will ignore this issue. However, problems that are inherently interesting to most pupils are likely to be more useful tools for teaching problem solving than those that aren't.

For most students there is really no particular obstacle to reaching the goal in usual school "word problems." Typically, they have just been taught precisely how to reach the goal. Indeed, when teachers fail to teach pupils precisely how to reach the goals, the pupils feel cheated—"You didn't show us how to do that problem." If the teacher had showed them "how to do that problem" it would no longer have been a problem. For those students who didn't understand the explanation, or who can't figure out which problem is to be done with which method of solution, or who were absent (physically or mentally) when the explanation was given, or who, for some other reason, don't already know how to answer the question, the "problem" is also not really a problem, it is simply a very frustrating situation in which they are doomed to fail.

Unfortunately, good problems are hard to create. Even if a teacher or textbook author or test writer creates a great problem, it is likely not to be a problem for a large share of the desired audience. The virtual certainty of failing for much of the potential audience should not, however, deter us from trying.

Examples of Good Problems

To give the flavor of a good problem, I now set forth several examples of situations that are likely to be good problems for at least some students. In choosing textbooks and tests, decision makers should look for these kinds of situations. Teachers should be encouraged to create as many such situations as they can beyond what is provided in textbooks and tests.

Games. Games like Cube–15, described earlier, offer excellent problem-solving opportunities. One of the benefits of such games is that some pupils will simply be practicing necessary skills while others will have proceeded on to some serious problem solving. Two children playing such a game together may be thinking on very different levels. However, it is likely that the player who is solving some serious problems will influence the other player after a while, either by action or by word. One of the experiences any teacher will have who uses such games is that somebody will accuse somebody else of cheating: "Kevin is looking at the cubes before rolling them." "It's not fair to think—this is math class."

The process of recognizing that there is a problem to be solved (How can I play this game so as to have a better chance of winning? Which cube should I roll next? Should I roll another cube or stop? Should my strategy change depending on the other player's score? Should my strategy change depending on what I think of the other player's ability? and so on) and expressing it in a form that helps to reach a solution is a very important part of problem solving. Games are one good way to practice this process.

Good, fun games that provide both worthwhile practice and serious problem-solving opportunities are not easy to create, and overworked teachers are not likely to produce many of them. However, there is no reason why a textbook should not provide good games of this sort to practice virtually every skill that children ought to practice. Integration of games that encourage thinking and problem solving into the textbook at appropriate places (so that children practice skills only after they have understood the underlying concept) should be considered a positive attribute of a textbook series. Textbook-adoption committees should look for such games.

Story Problems. Special stories that involve serious problem solving may be of help if written properly and treated effectively in the class. For younger children, such stories should probably be read by the teacher to the class to avoid issues of readability and to help keep children on task. Such stories should mix questions that are simple and straightforward with others that may require considerable thought or that may not even have one simple, correct answer. They should not, of course, rely heavily on the most recently taught skills since that reduces the need to think. Children should discuss these questions with the entire class.

For older children, such stories are more appropriately written in the pupils' books, and the stories should be read and solved in small groups (each group should include at least one good reader). Each small group should agree on its answers to the various questions and report back to the class (orally, in writing, or by other means such as drawings or pantomimes). Then, when appropriate, various groups may wish to debate the merits of the various solutions.

There is an important difference between these special stories and the activities in certain textbooks that may seem similar on the surface, but are essentially different. When my son was in 2nd grade he asked me for help with his homework one day. When we were finished, I noticed that the only interesting question on the page was in a box in the lower right hand corner. The box was labeled "THINK!" I pointed to the box and said, "Look, Todd, there's an interesting problem. Let's do that one."

He took one look at the word "THINK!" and responded, "We don't have to do that."

Although the problem was good, the labeling technique was terrible. All members of the class had discovered that they were never expected to think except when they were doing problems in such boxes. Furthermore, only the "better" students were ever asked to do such problems. Thus, nobody was ever challenged to decide whether a particular problem could be done by pedestrian procedures or required serious thought, and only the "better" students were expected to work on the nonpedestrian problems. But one of the most important parts of problem solving is deciding whether there is a problem in the first place, and the people who are most in need of practicing problem solving are the people who aren't naturally good at it. Thus, good nonroutine problems like the one Todd chose not to do should be mixed in with the regular problems on the page with no special symbols to set them apart as the problems to be done only by the good students when they want to think.

Good story problems can be pure fantasy (which can be quite real to most children and many adults), can be based on historical fact (for example, how did young Karl Gauss determine the sum of the numbers from 1 to 100 in a matter of a minute or two when his teacher assigned the problem to his class?), or may involve realistic situations (a realistic story about reducing air pollution or heat loss, for example).

Projects. Projects are often good ways to get children to do some serious problem solving. For example, which grocery store in the neighborhood gives the best buys?

Children usually start this project by assuming they can read advertisements in local papers to reach a conclusion, but then quickly decide that ads are likely to provide a biased sample. A grocery list is often unrealistic when made up by the children; but with parents' help, a fairly reasonable class, or group, grocery list usually can be developed. There are additional difficulties associated with different brands at different stores, differences in quality (especially of produce and meat), convenience, service, and so on

Usually children decide they can't really answer the original question, but they can collect a lot of interesting information that gives them strong arguments for one or the other of the stores and that might influence some shopping decisions. That's the way the world really is. Many questions that can be asked don't have simple answers; but we can often use mathematics to help us answer some of those questions, or at least to get a better understanding of the situation.

A project that will provide good applications of ratio is to have children measure various body parts, determine ratios, and then try to predict measurements for an unmeasured person based on one or two measurements and the average ratios for the class. In *Gulliver's Travels*, Part III, the tailors made suits by measuring only one body part (the thumb) and then using appropriate ratios to calculate all other dimensions. Gulliver remarked on the fact that nobody's suit seemed to fit very well, but failed to appreciate the remarkable fact that they fit at all.

Predictions made on the basis of such measurements and calculations will be less than perfect, but they will be surprisingly close in most cases. On the other hand, if the average ratio of head circumference to height for a 3rd grade class is used to predict the height of an adult (given the adult's head circumference) the prediction will be much further off than predictions for children of about the same age because this ratio changes as we mature physically.

Other potential projects are all around us. Have a class try to figure out how to park more cars, conveniently and safely, in the school parking lot, reduce the waiting time in the school lunch line without extending the lunch period too much, synchronize traffic lights on a local two-way street for a reasonable speed in both directions, figure out how to gerrymander the state so one particular party will get an overwhelming majority of the congressional seats—then do the same for the other party, and finally work out a fair redistricting arrangement, and so on.

A major deterrent to doing this sort of project is that teachers, parents, and pupils have a hard time categorizing such projects as clearly arithmetic, algebra, geometry, probability, or whatever. Most worthwhile applications of mathematics, however, are hard to categorize. This is not a good argument against such projects and applications—but is a rather good argument against the compartmentalization of school mathematics.

Applications to Mathematical Topics. There is a danger that when we speak of applying mathematics or of deriving the mathematics from the learner's reality we will fail to remember that many of the best applications of mathematics are to mathematics itself and that, after some years of studying, the learner's reality includes mathematics. Thus, it is perfectly reasonable to use a four-dimensional space as part of the learner's reality when teaching a course in complex variables even though four-dimensional space would seem quite abstract to many people. Similarly, the arithmetic of whole numbers may be real to a youngster learning algebra even though it was quite abstract to the same person several years earlier.

Consider the following example:

A student examines a table of squares and notices a peculiar pattern. The differences between successive squares seem to equal the sums of the numbers of which they are squares:

Number	1	2	3	4	5	6	7	8	9	10	11	12	13...
Square	1	4	9	16	25	36	49	64	81	100	121	144	169...
		3	5										25

Notice: $4 - 1 = 3$, and $1 + 2 = 3$; $9 - 4 = 5$, and $2 + 3 = 5$; ... $169 - 144 = 25$, and $12 + 13 = 25$. Will this go on forever? If so, how can we be sure? If not, when will it end? How can we test the conjecture about successive squares? We could try random examples to see if it works for them:

- $43^2 = 1,849$; $44^2 = 1,936$; $1,936 - 1,849 = 87$; $43 + 44 = 87$.

- $148^2 = 21,904$; $149^2 = 22,201$; $22,201 - 21,904 = 297$; $148 + 149 = 297$.

- $3,054^2 = 9,326,916$; $3,055^2 = 9,333,025$; $9,333,025 - 9,326,916 = 6,109$; $3,054 + 3,055 = 6,109$.

This looks very promising—but maybe we've chosen numbers for which it happens to work, and there are others for which it doesn't work. Shall we go on trying different numbers? If we tried a million different pairs of numbers and it worked for all of them, would we know for sure that it always works? If we tried a million different pairs of numbers and it failed to work for one of those pairs, would we know for sure whether it works for all pairs of numbers? (Yes. We'd know for sure that it doesn't, but we might be inclined to believe that it works for all pairs except that one.)

Let's let N stand for any whole number. Then $N + 1$ would be the next whole number after N. What is the relationship between N^2 and $(N + 1)^2$? If we know a little algebra, we know that $(N + 1)^2 = N^2 + 2N + 1$. The difference between that and N^2, of course, is just $2N + 1$, or $N + (N + 1)$, which is what we were trying to prove. So, we now know that the statement is true for all whole numbers.

Is this important? Is it interesting? Is it even worth remembering? Could we use it for anything? Let's try. Suppose we know that $40^2 = 1,600$ and would like to know what 41^2 is. How can we find out? We know that $41^2 = 40^2 + 40 + 41$, we can add 40 and 41 "in our heads," and we can square 40 in our heads. So, we can find the square of 41 by adding 1,600 and 81, getting 1,681. Similarly, 91^2 is $8,100 + 181$ or $8,281$, and so on.

Also, a bit of thought will convince us that 89^2 is $8,100 - (90 + 89)$ or $8,100 - 179$ or $7,921$. So we can use this theorem to calculate mentally the squares of numbers that happen to be near (one away from) any whole number for which we already know the square. That seems reasonably useful for some purposes.

As well as reviewing the theorem itself to see if it is sufficiently useful or interesting to remember and to use in the future, you might also consider whether the process through which we went to discover and prove it are generalizable and likely to be useful. Are there other interesting, testable patterns? Does letting a letter stand for some general number make sense, and is it helpful? Is algebra useful in proving general statements about arithmetic?

The particular trick shown here is not of great importance. But there are several important messages. First, mathematics can be applied, and should be applied, to problems within mathematics as well as outside mathematics. Second, by thinking mathematically, people can make life easier for themselves. There are many instances in which a little bit of algebra can make arithmetic computations much easier, and similar examples abound in other areas of mathematics as well. Third, by compartmentalizing mathematics, we are likely to overlook many of these examples. We should therefore do all we can to decompartmentalize school mathematics (this issue is pursued further in Chapter 6, "Connections").

One further comment about the relationship of algebra to arithmetic: by the time students start studying algebra, they usually are so familiar with arithmetic that it is part of their reality. In that case, the algebra should be developed, at least partially, from that reality.

For example, if a teacher is teaching how to multiply $(3x + 5)$ times $(2x + 7)$, arithmetic examples might come first. Try multiplying 35 times 27, for example. We would multiply 5 times 7, getting 35; then we would multiply 5 times 2, getting 10 (tens), and add, getting $100 + 35$. Next we would multiply 3 (tens) times 7 and then times 2 tens, and finally we would add the four partial products together.

$$
\begin{array}{r}
27 \\
\times 35 \\
\hline
35 \\
100 \\
210 \\
600 \\
\hline
945
\end{array}
$$

A teacher could have students work this out in groups and keep careful track of what they are actually doing, and then, by analogy, try to work out the problem involving $3x + 5$ and $2x + 7$, thinking of the x as 10. After several experiences like that, the students should have a fairly clear idea of the relationship between the arithmetic examples and the algebra.

Space Applications. The most common applications of mathematics to the world around us involve number and space. For some reason, we tend to neglect the space applications in school mathematics—and when we do study geometry, it is usually two-dimensional (or plane) geometry.

We live in a three-dimensional world and constantly have to make decisions about that three-dimensional world on the basis of available information. We see only a two-dimensional world—or more accurately, we see two two-dimensional worlds, one with each eye. Our brains make inferences about the three-dimensional world from the two, slightly different, two-dimensional pictures. Young children ought to have lots of experiences that emphasize this relationship and that help the children understand both two- and three-dimensional space, as well as the relationship between them.

Such experiences might include predicting how many times they can fill a small container and pour its contents into a large container. They would then experiment to get an actual count (usually more often than people predict). They could repeat this activity using a conical container and a cylindrical container of the same height and base.

At a later time they could derive formulas for volume and see why their intuition led them astray. The volume of similar containers varies as the cube of a length, but the area we see varies as the square of a length. Thus, doubling a length multiplies the volume of a container by 8 while only multiplying the area we see by 4. Therefore, the smaller container seems larger with respect to the bigger container than it actually is. Similarly, a cone has a volume one-third the corresponding cylinder, but the area (of the triangle) we see for the cone is half the area (of the rectangle) we see for the cylinder.

Such discussions can be used to explain the surprising volumes of various containers, and for many other purposes. For example, J.B.S. Haldane (1985) used this kind of analysis in a book called *On Being the Right Size* to show why two closely related animals are such different shapes. As height (say) is doubled, volume is multiplied by eight and area by only four. Cross-sections of legs must expand disproportionately quickly (to withstand the increased mass), as must areas of exposed skin (to cool the body and take in oxygen).

Very young children (five or six years old) can make scale models of their classrooms and use them to see what would happen if they moved furniture around, moved themselves about the room, or otherwise changed things. From the scale model they can make scale drawings (by looking straight down on the model). Later, they can study maps, blueprints, and other two-dimensional models of the three-dimensional world.

Somewhat older children can use paper folding to do various two-dimensional activities, such as constructing perpendicular bisectors and angle bisectors, and can also fold two-dimensional paper to make three-dimensional objects. By the age of 12 or 13, children are able to construct all the regular solids and give a convincing argument that there can be no more than five of them (see also Chapter 6, "Connections," for a further discussion of regular solids).

Lots of physical experiences with both two- and three- dimensional geometry should be provided in school. Those experiences should be examined abstractly so that children get a good feeling for two- and three-dimensional space—and also learn that by thinking abstractly about space they can reach conclusions that might have escaped them if they had limited themselves only to physical experience.

Several reasons are usually given to justify the traditional 10th grade course in Euclidean geometry. One of these is to provide a better understanding of space. In fact, when the appropriate physical experiences have been provided in previous grades, there is very little more that children are likely to learn about space in 10th grade geometry, other than some relations involving lines and circles (e.g., inscribed angles, tangents, secants, and chords, and angles involving them). With the general lack of attention to three-dimensional space in such courses, there is little likelihood that they will learn much that is new and worthwhile.

A second goal often given for the traditional 10th grade geometry course is to teach the students the nature of proof. Unfortunately, few learn much about proof. Many children apparently are simply not sufficiently mature to appreciate formal proofs, and even if they were, they would get a distorted notion of proof from the typical two-column, statement-reason type of proof offered in traditional geometry courses. Outside of such courses, two-column proofs are never used by anybody. Mathematicians, scientists, attorneys, politicians, teachers, philosophers, and other people, including children, who wish to give convincing arguments use whatever form of proof that appeals to them, including paragraph arguments, pictures, appeal to authority, and so on.

In the early 1960s, Edith Robinson (1964) interviewed a group of high school sophomores, half of whom had taken plane geometry and half of whom had not. They were all convinced that a particular geometric fact was true. They were asked to give a convincing argument to her. Independent of whether they had had a course in geometry, not one of them gave a statement-reason proof.

Even if statement-reason-type proofs were widely used, the usual procedure in geometry courses is to tell students to prove many different theorems in essentially one way. How much more educational it would be if one theorem were proved in many different ways. Indeed, with all problem solving, there is generally more to be gained by solving a single problem in several ways than by solving several problems in a single way.

Deductive Proof

To many people, logical proof is at the heart of mathematics. Many cultures had developed a great deal of very good mathematics before the ancient Greeks. A major contribution of the Greeks was to take many different mathematical statements and put them together in a way that showed their logical relationship to each other. Insofar as you believe in those proofs, and you believe the very small number of assumptions (or axioms) made by Euclid, you are required to believe all of Euclid's theorems. So, if you can convince yourself of the validity of the very few axioms, you can safely believe in all the theorems. For 2,000 years, Euclid was considered by educated people to be the epitome of logical argument.

Our understanding of proof and structure have changed over the centuries, but Euclid is still a remarkably good example of both.

Ideally, a mathematician starts with a small number of axioms or assumptions about a situation and then derives or proves theorems. If the proofs are correct, then the theorems are true whenever the assumptions are true. Supposedly, the initial axioms need not apply to any particular situation. They may or may not be true of some particular part of the real world. In practice, most mathematical systems that are taken seriously have their genesis either in the real world or in some theoretical consideration (such as mathematics itself).

Applied mathematicians can save a great deal of expense in time, money, and danger by constructing a mathematical model of a situation and predicting what will happen in advance. If their logic and assumptions are correct, it may not be necessary to orbit an astronaut, try a drug on thousands of sick people, or perform some other experiment because the theorems proved will predict accurately what will happen. Of course, scientists must constantly have reality checks to be sure the assumptions and logic have not deviated too far from reality.

Mathematical systems and proofs can also be used to predict things that nobody would ever have imagined without the mathematics. By deriving some unlikely looking theorems, a mathematician may call attention to a logical result that seems to defy common sense. As a simple example, suppose you are in a spaceship chasing a space telescope. You are both in the same orbit, but you are ten miles behind. How can you catch up, with the least expenditure of rocket fuel? Mathematics will tell you (correctly) that you should decelerate! By doing so, you will drop into a smaller orbit and will have greater angular velocity than before. When you have caught up with the telescope you should accelerate enough to get out into the telescope's orbit. "Common sense" is unlikely to lead you to this conclusion. A mathematical model based on logical proofs and a few correct assumptions will.

Part of the power of mathematics comes from the fact that a given system may be applicable to many quite different situations. In the case of the spaceship and the telescope, there is a well-developed mathematical system that describes motion of various bodies in space. Newton used the system to describe the motion of the planets and various other heavenly bodies, motion on earth, and various other matters of interest. Quite often, the axioms of a given mathematical system turn out to be applicable to situations that seem totally unrelated. In such a case, all the theorems proved about the first situation turn out to be true about the new situation, without the need of reproving each of those theorems.

Mathematical models or systems have made possible the creation of our highly technological world. They are applied in medicine, economics, physics, politics, psychology, education, and a host of other human endeavors. Proof is central to such mathematical models, because proof is needed to show that certain results (theorems) follow from certain assumptions (axioms) or other theorems.

As is often true with really important ideas, there is a good deal of disagreement among "experts" as to precisely what a proof really is. A proof is nothing more—nor less—than an argument that convinces somebody. What is acceptable as a proof changes depending on when the proof is given, who is supposed to be convinced, what the nature of the question is, and many other variables. To try to teach people how to construct a proof independent of these many variables is silly.

In general, when we try to teach people about proof, we should also give them experience with the activities that usually relate to proofs. In real life, we seldom try to prove something until we have convinced ourselves that it is probably true. We don't usually convince ourselves until we have experimented with various alternatives and chosen what appears to be the most viable one. In school, too, we should encourage children to think about a particular subject, experiment, speculate, conjecture, and test before they set out to try to prove something about it. This is the procedure we followed in the squares of numbers problem discussed earlier in this chapter. It can also be applied to geometry.

For example, draw a circle (Figure 4.1). Draw a line through the center, O, of the circle, cutting the circle at points P and Q (so that PQ is

FIGURE 4.1

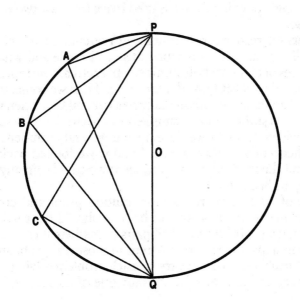

a diameter of the circle). Now draw some angles whose sides go through P and Q respectively and whose vertices are on the circle: angles PAQ, PBQ, and PCQ, for example. Do you notice anything interesting about all the angles? Did you expect them all to be right angles before you drew the picture? Do you believe they really are, or did you just draw a few that happened to turn out that way? Do you think the same thing would be true of other angles drawn the same way? Measure each of these angles and several more that you draw to see if you can find one of these angles for which your conjecture is not true. If the angle measure is just a little different from 90 degrees, do you think this is likely to be because the angle is not really a right angle or because you've made a slight error in drawing or measuring?

Now, try to write a convincing argument that your conjecture is true. Such an argument may require you to use a theorem that says that an angle inscribed in a circle is measured by half the intercepted arc. You may have to prove that theorem for your disbelieving audience by drawing such an angle and a central angle intersecting the circle in the same points, and so on. The argument, however, should depend on both the facts of the situation and on the sophistication of the audience.

Both you and the audience may now wish to consider whether this particular bit of information is interesting enough, or is likely to be useful enough, to have been worth the effort to discover and prove it, and whether it is worth remembering now that you know it. You may also wish to consider whether there is anything general about either the theorem, or the process we used to discover and prove it, that might be useful or interesting for future problems. Incidentally, the converse of this theorem can be (and often is) used to find the center of a given circle. A T-square or other right-angle drawing device is used twice (to draw two diameters) for this purpose.

The observant reader can hardly avoid noticing a parallel between what we just did for an angle inscribed in a semi-circle and what we did earlier for squares of certain whole numbers. In each case we experimented and noticed something that looked interesting. Then we conjectured that the interesting "coincidence" might really be a general fact. Then we tested the conjecture with several more examples until we were pretty sure it was not just a coincidence. Then we developed a general argument that the conjectured theorem was always true. Finally, we looked back at both the theorem and the method we used to see if we could gain any insights that might help us in the future.

There is, of course, no reason for limiting proof and concurrent activities to geometry. Such work can be, and should be, carried on in algebra, arithmetic, probability, and all parts of mathematics.

These same steps, in fact, are inherent in most real problem-solving situations. Unfortunately, when we try to teach problem solving and proof in school, we tend to limit ourselves to only one of the steps of this very rich process—the formal derivation or proof. We seldom or never let the students investigate and identify their own problems and theorems, and then test their conjectures in the absence of an authority who will tell them the conjecture is true or false. We seldom if ever encourage them to define their own audience and try to give a convincing argument for *that* audience. We trivialize the entire process in the hope that this will somehow make students more likely to understand. Instead, they are less likely to understand, and they are almost certain not to find the process worth emulating.

Importance of Communication

Communication is, in many respects, the most important part of mathematics. In an axiomatic system, all of the information is inherent in the original axioms. Theorems, in general, contain less information than (or, at most, the same amount of information as) the original set of axioms. The reason for stating and proving the theorems is that the information is more useful, or easier to comprehend in that form. Thus, the activity of stating and proving theorems is essentially one of communicating.

In a similar way, a mathematician who works for industry may be asked a question and may find an answer to the question. The fact that the mathematician knows an answer is of absolutely no interest unless the mathematician can communicate the answer to someone who will use it. Furthermore, unless the mathematician can understand the original question when it is expressed (often in less than completely cogent terms) by the original poser of the problem, the mathematician is of little use to the industry.

Communication is, and always has been, an important part of mathematical problem solving. Recent advances in technology have made certain activities (computation, solving equations, and other symbol manipulating) more appropriately done by machine than by the human mind. However, use of such machines requires the mathematician to be able to communicate with machines as well as people. Thus, communication has become even more important than before.

Communication may be oral, it may be written, or it may take other forms such as building a model or drawing a picture. But mathematicians must be able to receive and understand communications, and they must be able to communicate back their results.

In school mathematics, we have a tendency to slight this all-important aspect of mathematics. We almost always communicate the problem to be solved or the theorem to be proved in writing, sometimes with words, often only with mathematical symbols.

We accept answers like "73" or "53 meters" that could not, by any stretch of the imagination, be called English sentences. Almost never does a mathematics teacher require students to produce answers to questions in well-written English sentences that are grammatically correct. Almost never do we encourage students to communicate with each other, using language or other means, to express their discoveries and beliefs about mathematics. We are simply remiss about teaching children to communicate effectively, and by being remiss, we are doing considerable damage to those students; indeed, we are reducing their opportunity to make significant contributions to the future of the world, or even to understand such contributions when they are made by others.

Student-Formulated Problems

An effective method of encouraging children to think about written "word problems" is to have them make up some of their own. A general assignment to "Make up your own problems," however, usually produces unpleasant noises and few good problems. On the other hand, if the teacher brings in a newspaper (or news magazine), tears off a page for each group of three or four children, and says "make up three problems based on information you find on either side of that page," very good problems usually result.

There is almost always enough information for more than three good problems. There is usually so much information that I discourage the groups from making up problems based on advertisements because that's too easy. Children often discover that there are some statements in the article that seem to suggest a conclusion that can't be derived on the basis of the available information. This encourages children to read such news articles much more carefully thereafter. Such activity also emphasizes the connections between mathematics and reading and writing, as well as social studies and other subjects of the articles.

Occasionally, children will discover articles that provide too much information—that is, from the information in the article you can derive two different answers to the same problem. Articles with contradictory information don't occur very often, but they are certainly worth examining when they do.

The real goal of having students make up their own word problems is to encourage them to think seriously about the meaning of words and sentences as they relate to real-world mathematical material. Going through the process of making up such problems, even as a member of a group, usually helps children read and understand such problems when they are created by others.

Of course, the group making up a problem is expected to determine the answer (which may be that there is too little or too much information given). Then the groups exchange problem sets and check to see if each group got the same answers for the problems.

Activities of this sort turn out to be one of the best ways to help children become better solvers of word problems, and such sessions should probably occur several times a year.

Problem-Solving Strategies

Telling children strategies or rules for solving problems in advance is not an effective way of improving their problem-solving skills, though it is a way to help fool ourselves and others into believing we have helped them. This is particularly true if a strategy is supplied and then a bunch of problems, all of which use that one strategy, follow. In such cases, the pupil is simply learning by rote to apply the particular process to any problem that is on the appropriate page. This is of little or no help, since the real issue is deciding which process is appropriate. For this reason, problem-solving rules in general are not especially helpful and may have a negative effect if they are introduced one at a time with numerous examples using each strategy as it is introduced.

On the other hand, it is well established that we should look back at the processes we use to solve problems, think more deeply about those processes, and even verbalize our thoughts to help us decide when the processes are likely to be useful in the future.

By the time students are in junior high school, many have become sufficiently mature to think seriously about their own thinking process (this is called metacognition in some circles). If they have had sufficient experience solving good mathematical problems, a list of general strategies for them to consider, reject if they choose, modify if they wish, but at least to think about, may be helpful.

Many people have produced such lists of problem-solving strategies. Perhaps the most famous of these is the set produced by George Polya (1957) in his book *How to Solve It*.

Students should be encouraged to produce their own lists of strategies, to think about their own and other people's strategies, and then to continue solving many interesting and difficult problems using whatever strategies they wish. But after solving such problems, they should be encouraged to rethink and discuss what they have done, decide which strategies have been useful and which have not, and continually reexamine their own and other people's methods of solving problems. They should be encouraged to discuss their lists with each other and modify their lists in light of suggestions by others. No single list of strategies should be taken very seriously except the student's own list, and that list should always be subject to modification.

Cooperative Learning

From time to time, education goes through various phases, or fads. The creators and supporters of these fads always come up with attractive-sounding names: "discovery approach," "individualized instruction," "mastery learning," "cooperative learning," and so on. The practices that then masquerade under the attractive-sounding titles are not always so attractive. Much of what went on under the guise of individualized instruction, for example, seemed more like solitary confinement to some of the victims. For that reason, I prefer to describe in some detail what I think should and shouldn't happen with learners rather than to simply subscribe to a particular new phrase or phase.

No one who has read this far can possibly fail to notice that every aspect of learning mathematics that I have described involved learners working together. That's the way people do mathematics. That's the way people should learn mathematics.

If the mathematics department of a major university wishes to hire a point set topologist and has none at present, it will have to hire two of them. One would not be willing to work in isolation without someone to talk with—to try new ideas on, to check each other's reasoning, and so on.

If I plan to cook two roast beefs that weigh 10 pounds each in the same oven, and the cookbook suggests I will need 20 minutes per pound at the chosen temperature, I would want to discuss with somebody who knows more than I do about such things whether the correct time is more likely to be 200 minutes or 400 minutes. I suspect the timing depends partially on what kind of oven I'm using (microwave or conventional). Whenever we do mathematics in situations that are important to us, most of us try to discuss it with somebody else.

Why, then, do we try to make children learn mathematics in isolation? Mathematics is not a solitary activity. It should be done and learned with others. Games, activities, projects, proofs, problem-formulation activities, and so on are all activities that should be carried on in groups. Where possible, those groups should involve children of different abilities, different interests, and different backgrounds; and each member of the group should be expected to make substantial contributions and derive substantial satisfaction. This was true ten years ago, it is true today, and it will be true in ten years—whether or not "cooperative learning" happens to be in vogue.

The answer to the bear problem presented earlier in this chapter is "white," since the hunter presumably started at the North Pole. There are, however, other locations from which the hunter could have started and followed a path like the one described (though she'd be unlikely to meet a bear). Consider a circle with its center at the South Pole and a one-mile circumference. Start one mile north of that circle. Replacing the circle with one of circumference 1/2, 1/3, or 1/N miles, where N is a natural number, will also work (but the hunter may become dizzy).

References and Further Resources

Burns, M. (1976). *The Book of Think: Or How to Solve Problems Twice Your Size*. Boston: Little, Brown.

Burns, M., and B. Tank. (1988). *A Collection of Math Lessons from Grades 1 Through 3*. New Rochelle, N.Y.: Math Solutions Publications.

Erlwanger, S.H. (1974). "Case Studies of Children's Conception of Mathematics." Doctoral diss., University of Illinois, Urbana-Champaigne.

Haldane, J.B.S. (1985). *On Being the Right Size and Other Essays*. (Reprinted). Oxford: Oxford University Press.

National Council of Teachers of Mathematics. (1989). *Curriculum and Evaluation Standards for School Mathematics*. Reston, Va.: NCTM.

Polya, G. (1957). *How to Solve It: A New Aspect of Mathematical Method*. 2nd ed. New York: Doubleday.

Robinson, G. (1964). "An Investigation of Junior High School Students' Spontaneous Use of Proof to Justify Mathematical Generalizations." Doctoral diss., University of Wisconsin, Madison.

Stevenson, P.R. (1925). "Difficulties in Problem Solving." *Journal of Educational Research* 11: 95–103.

Willoughby, S.S., C. Bereiter, P. Hilton, and J.H Rubinstein. (1981, 1985, 1987). *Real Math, K–8 Textbook Series*. La Salle, Ill.: Open Court.

5

Technological Developments

Rapid developments in technology are changing (and ought to be changing) the way we teach mathematics both because they modify our goals for the mathematics education of people and because they provide new tools with which we can better achieve our goals.

Thinking and Technology

Calculators and computers are here to stay and will continue to become more useful and easier to use. Conversely, it seems to be true that people will always be able to do certain things that machines can't do and should be educated to do those things well rather than being trained to do what a cheap calculator or a computer can do better.

Currently, there are calculators and computers costing less than $100 that can perform most of the mathematical symbol manipulation taught in schools between kindergarten and the second year of calculus. These machines can do arithmetic with whole numbers, rational numbers, complex numbers, and vectors. They can solve equations and systems of equations; they perform graphing functions and can "zoom" in on parts of the graph to get a magnified picture. Machines can perform algebraic differentiation, integration of functions, and most other mathematical symbol manipulations taught in schools and colleges.

Does this mean that people no longer need to learn to do arithmetic, algebra, calculus, and so on? Of course not. Learning mathematics is more important than ever, but the specific skills needed are changing.

Nor is there reason to believe that people will no longer need to manipulate symbols. The ability to do certain kinds of symbol manipulations or perform certain rote skills efficiently helps people perform many of the higher-order tasks. For example, knowledge of the "number facts" (addition to $10 + 10$, subtraction to $20 - 10$, and so on) is essential to mental calculations and estimations. The ability to estimate allows us to use calculators and computers more intelligently—catching obvious errors

and even avoiding use of the machines when our minds can do the task more readily.

However, rather than simply continuing to teach students all the skills and knowledge that have been assumed to be useful in the past, we must now decide which abilities people of the next century will need and which they will not need. There is a limited amount of time for educating people, and we cannot afford to waste any of it teaching things that are certainly going to be useless. On the other hand, we must not forget to teach them the more mundane skills that are prerequisite to the higher-order skills, and we must be careful not to overlook relatively pedestrian skills that we now take for granted.

People must learn to do well those things that they can do better than machines. Such skills include problem identification and formulation, and restructuring a problem into a form in which it can be solved with available tools (including, but not limited to, computers and calculators). People must develop their abilities to:

- choose the most appropriate available intellectual and mechanical tools to solve a problem,
- decide when to give up (because the problem can't be solved, or is so messy to solve with available tools that it is not worth solving),
- recognize reasonable and unreasonable solutions,
- interpret the solution of a problem reasonably, and
- reexamine a solved problem to look for better solutions, generalize the solution, and consider other related problems that are interesting and might be solved in a similar manner.

Perhaps the most important skills the schools can teach are flexibility and the ability and inclination to learn new things in the future.

Calculators in School and Society

Several years ago the mathematics supervisor of a small New England city recommended that calculators be used in the elementary schools so that pupils could be taught to use them intelligently and also to allow more interesting mathematics to be done in the schools. The school board reviewed this proposal and passed a resolution saying that children were not to be allowed to have or use calculators of any sort in schools. Furthermore, no teachers or administrators were to use or to be seen in possession of calculators in the schools. But to show that they were not against progress, the members of the school board also appropriated enough money to purchase several computers for the elementary schools. The computers were to be used to train children to do precisely those things that a $5 calculator can do better than people—multidigit arithmetic problems and other mechanical symbol manipulation.

Unfortunately, this is not an unusual case. For some reason, the public at large has gotten the strange notion that using a calculator in school is somehow cheating, but using a computer is good. The same people who object to the use of calculators in school usually use (or probably should use) calculators themselves in their everyday activities. And usually they use those calculators far less intelligently and efficiently than they would if they had been taught to use them well.

Some of the most commonly heard arguments against using calculators in school are: (1) children will not learn the necessary skills if they are allowed to use calculators, (2) people will become so dependent on a calculator that they can do nothing without the calculator ("What will they do if the battery fails?"), (3) the use of calculators is inherently unfair because rich families will be able to buy better calculators for their children than poor families, and (4) administrative care of the calculators is too difficult (passing them out, collecting them, making sure each child has a working calculator, preventing theft, and so on).

There is now substantial research evidence available that shows that use of calculators does not interfere with the learning of necessary skills. Numerous studies have been conducted at all levels and have shown that the use of calculators does not interfere with the learning of basic number facts, with other forms of computation, or with the learning of any other skill that is commonly thought to be useful. The one exception to this general statement is at the 4th grade level, where there is some slight evidence that the use of calculators may interfere with the acquisition of certain numerical skills usually tested at the end of grade 4. This appears to be an anomaly resulting from the content of 4th grade standardized tests.

The argument about becoming dependent on calculators seems reasonable until it is analyzed more carefully. Beyond the evidence that there is no substantial loss of skills among children using calculators, there is the question of whether we really want children to be able to duplicate (more slowly and less accurately) what a $5 calculator does well. When automobiles first appeared, there were undoubtedly many people who kept a spare horse in the garage lest the automobile fail, but very few people do so today. Calculators, with and without batteries, have become so inexpensive and reliable that it is more efficient to keep an extra calculator handy than it is to learn to do well everything a calculator does better. Most of us no longer find the ability to shoe a horse and cinch a saddle to be essential skills. Is it not reasonable to suppose that in the near future we may feel the same way about multidigit long division?

The equity argument against the use of calculators in schools and on tests may have been valid for several years during the 1970s when calculators were quite expensive and there was a great variation among calculators. Today, good calculators can be bought for less than $5, and sophisticated scientific calculators are commonly sold for less than $25. In spite of this, the problem could be serious if the ability to purchase a sophisticated calculator could really make a difference in a person's score on a college entrance examination or in a school course. There are ways to solve this problem, such as controlling the kinds of calculators that can be used or asking questions that will not be easier for a person with a more sophisticated calculator; but teachers and test makers should probably keep the argument in mind, not as a reason for not allowing the use of calculators, but rather as a caution in deciding what kinds of calculators to allow and what kinds of questions to ask on tests.

A report of research done by Gary Bitter (1989) of Arizona State University raises a quite different question of equity. He reported that girls, who have traditionally performed more poorly in mathematics than boys, scored as well or better than boys after using calculators in mathematics class for a year, and that even boys using calculators did better than the boys who had not used them. If this result is duplicated in further research, and if the same results turn out to be true for other traditionally underrepresented groups in mathematics, the equity argument for using calculators in school mathematics will be irresistible.

When calculators were expensive, rare, and very primitive, the control and maintenance of them in a classroom was a problem, but one that could be handled. I used them in a class of 37 5th graders in 1978. We lost no calculators to theft, and only one to an accident. Protecting batteries was one of the most serious problems then. If a child inadvertently left a calculator turned on, the battery would be dead the next day. Today's calculators either are solar powered or turn themselves off after a short inactive period. Besides that, the calculators take so little electricity and the batteries are so good that battery-powered calculators can be left on for several years before the battery wears out. Today, the administrative problems associated with using calculators should be no more complex than those associated with using textbooks, pencils, papers, and other common school materials.

The fact that most of the arguments against using calculators in the schools don't stand up under careful scrutiny does not, of course, imply that they should be used in schools. Some clear benefit should accrue to the learners from using new procedures or materials to justify the expense and inconvenience of change.

The most obvious reason for teaching children to use calculators is that they are all around us in the world outside of school, and most people who have access to them do not use them very intelligently. Since calculators can be very powerful tools in doing mathematics, one of the obligations of a good mathematics education program in a school is to teach students how to use calculators intelligently, including when *not* to use them because there are better tools available for the task at hand. Beyond that, it is hard to convince children that school mathematics has something to do with the real world if they see everybody outside of school doing mathematics with calculators but they are not allowed to use them in school. Of course, the equity issue mentioned earlier may become the strongest argument for using calculators in school.

Calculators and the School Curriculum

Some changes ought to be made in the school curriculum to prepare children to use calculators more intelligently. Undoubtedly the most obvious and most essential change of this sort is to introduce decimals earlier. Because the United States does not use the metric system of measurement, as do all other industrialized countries, our curriculum introduces decimals later and places less emphasis on them after they are introduced, than do other countries. Whether or not the metric system is taught in schools, there is no reason why decimal fractions should not be introduced at least by grade 2.

The monetary system of the United States provides good motivation for introducing decimals; and if the metric system is used, it too can provide a good model for decimals, as can calculators and computers themselves. Decimal fractions and common fractions should be taught independently of each other for a while (at least until children can use calculators to convert common fractions to decimal fractions and probably until about grade 5 so that the connection can be understood). After this grade level, decimal and common fractions can be treated as essentially one topic. Many textbooks and curriculum guides provide good procedures for teaching children decimals.

The most obvious and important reason for teaching people to use calculators in school is that they will use them more intelligently if they are taught how. Intelligent use of a calculator requires skill with the basic number facts, knowledge of the base ten system, and good number sense. These are all things that should be learned in school as a standard practice. After calculators have been introduced and pupils have learned to use them reasonably efficiently, a game such as the following "race" will help promote the intelligent use of calculators.

Some of the children in the class will be allowed to use calculators, and others will not. The ones who are not going to be allowed to use calculators usually complain when they hear this.

To make the race seem fairer, the teacher announces that those with calculators will be required to push every key—not skipping any or doing anything "in their heads." That is, if the problem is 7 x 58, they must push: 7, x, 5, 8, =, without skipping any steps such as, for example, noting that 7 x 60 is 420 so the answer must be two 7's less than 420 or 406.

This usually seems like a very minor restriction, and the people with calculators are still happy and the people without them are usually still upset.

Next, have them write the problem numbers on their papers—a column from 1 to 15 down the left side of the paper and from 16 to 30 down the middle—so they won't be slowed down with any more writing than necessary.

Watch the people with calculators carefully to see that they don't skip steps as they do the problems in Figure 5.1

FIGURE 5.1

1. $10 \times 73 =$
2. $100 \times 73 =$
3. $1,000 \times 73 =$
4. $10,000 \times 73 =$
5. $100,000 \times 73 =$
6. $10 + 73 =$
7. $100 + 73 =$
8. $1,000 + 73 =$
9. $10,000 + 73 =$
10. $100,0000 + 73 =$
11. $800 + 500 =$
12. $800 - 500 =$
13. $8 \times 5 =$
14. $7,568 \times 0 =$
15. $84,595 + 0 =$
16. $730 \div 10 =$
17. $7,300 \div 100 =$
18. $73,000 \div 1000 =$
19. $730,000 \div 10,000 =$
20. $7,300,000 \div 100,000 =$
21. $4 + 8 =$
22. $7 + 9 =$
23. $9 - 7 =$
24. $10 \times 8 =$
25. $101 - 1 =$
26. $63 \div 7 =$
27. $56 \div 8 =$
28. $6 \times 8 =$
29. $7 \times 7 =$
30. $1,000,000,000 \times 10 =$

Usually more than half the students without calculators have finished the 30 problems correctly before the first person with a calculator finishes. The students understand the message with no further explanation.

In fact, of course, the race cannot be completed by a person using an ordinary calculator, following these rules, because problem 30 cannot be done. There are too many digits in the first factor.

To avoid the inference that calculators are always inefficient, pupils can be asked to carry on a similar race with exercises that involve messy computations, such as $34,902 + 6,935$.

The moral of all of this is that people will not necessarily learn to use calculators intelligently unless they are explicitly taught to do so. Students will first have to understand the base ten system and learn the basic "table" facts so they can do mental arithmetic easily and estimate efficiently. They must develop a good number sense and be able to evaluate the reasonableness of answers to use technology intelligently. Beyond that, they will have to become accustomed to thinking and making judgments when they do mathematics. Students should regularly use calculators in school, with appropriate instruction, so that they will learn to integrate their various skills to solve mathematical problems efficiently. To restrict calculators in school or to continue teaching the same topics in the same way (ignoring the technological changes taking place) would be an anti-intellectual, "head-in-the-sand" way of failing to prepare our children to live in the 21st century.

Computers in School

In some respects, calculators and computers are similar in the influence they should have on school mathematics. Both devices are designed to do the more pedestrian drudgery often associated with mathematics; and both require an intelligent, thoughtful operator if they are to be used efficiently. Neither calculators nor computers can compare with human beings in the higher-order thinking skills. As calculators have become steadily more sophisticated, it has become much more difficult to define or describe the difference between a calculator and a computer. Size, programmability, memory, price, ease of use, and other features have been used in the past to distinguish computers from calculators; but distinctions on these variables have become progressively harder to make as technology has developed.

Considering the similarities, the difference in public perception about the desirability of using calculators and computers in schools is amazing. Many otherwise intelligent, thoughtful people, including some educators, seem to have the strange notion that computers are good and calculators are bad, even though they do essentially the same things. Because of the greater sophistication of computers and because so many commercial programs are available, computers can probably be used to do more good—or more harm—in school than calculators.

Because of rapid changes in technology, any comment made now about the differences between calculators and computers is likely to be obsolete within a year or two. The reader should interpret any remarks in light of subsequent developments. At present, however, the most significant difference seems to be the availability of large numbers of very sophisticated programs for computers. There are word processing programs, spreadsheet programs, database programs, graphing programs, computational programs for doing the simplest and most complex symbol manipulations, game-playing programs, simulation programs, teaching programs, and all sorts of other programs, including virtually every combination of such programs that one can imagine.

Because of the availability of large numbers of computer programs, the user or potential user of computers must make some sort of evaluation of programs before purchasing (or otherwise deciding to use) a program. Even if a computer and a copy of all potential programs were available, evaluation is a major task requiring a great amount of time and knowledge. Most normal people are obliged to rely on, at least partially, on the judgment of others. You may take the word of a friend or sales representative. You may read reviews published in various professional journals. You may buy various commercial reviews that are available (either through computer modem services or in print).

Whatever procedure you use, there are several possible pitfalls. Almost certainly, shortly after you make your selection, new hardware or software will become available that will make your selection seem less than optimal. On the other hand, if you wait for the best possible combination of computer and program, you will never make a choice, so you might as well make the decision as soon as you are ready.

A far more serious problem is the possibility of choosing a program or computer that is essentially wrong. In choosing a business program or a program to schedule students, you might get one that doesn't do the job the way you want it done. For example, the needed information may not be retrievable or the program may not be flexible enough to do what really needs to be done.

The various kinds of difficulties people encounter choosing most business or administrative programs are both more obvious and relatively less important than the kinds of mistakes that can be made in choosing instructional programs. There are some delightful teaching programs that encourage creativity and higher-level thinking skills on the part of the user, but the majority of instructional programs in mathematics (more than 90 percent by one reliable estimate) have as their goal to train the user in some particular low-level skill. That is, the computer program is designed to use technology to teach the student to do something that is actually done better by a computer or calculator. This tends to be a waste of computer time and often of student time, even though it may occasionally have some beneficial effects.

Because many computer programs are written by people who are neither educators nor scholars, some have serious mistakes of fact or pedagogical strategy, or both. These may be hard to notice unless a competent person goes through the entire program in much the way the learner would be expected to do—a time-consuming, often boring process.

The really serious difficulty with such programs, however, is that they tend to leave a distorted idea of the relationship between humans and computers. Since the computer "knows" all the answers and seems to sit in judgment of the human, the student may begin to think of the computer as master and humans as servants. Many people in our society already behave this way. If the bank computer says your checking account has $1.17

less than you think it has, do you believe your own figures or the bank's? Most people will take the bank's figures, even though computers often give wrong information—sometimes because of keypunch errors, sometimes because they have been deliberately programmed to make small errors in the bank's favor, sometimes for other reasons.

Anything that schools do to reinforce the notion that computers should be believed over human beings is dangerous and should be avoided at all costs. Because of this, and the difficulty of evaluating computer teaching programs, and the expense of using them, such programs should be used with great caution. If the same goals can be achieved through printed material, flashcards, games, and activities, these alternative methods should be given careful consideration. When computer programs are used for instructional purposes, those that make the learner an active participant rather than a passive receptacle should be preferred, just as with other materials or teaching strategies.

In general, before choosing any computer program, you should decide what you want the program to do for you and write out in some detail the things you want and don't want the program to do. Then you should be reasonably certain that any reviewer you consult (whether friend, sales representative, professional reviewer, or other source) has the same goals in mind in making evaluations—or at least gives you sufficient information so that you can decide whether the program achieves those goals. Then, if possible, you should try one or two programs that seem most likely to meet your needs. In doing so, you may discover other characteristics of the programs that you had never thought of—some will be positive and some negative. Take these into consideration too, but don't lose sight of your original goals.

Ideally, choosing a computer should occur after you have decided on programs, since programs that are well suited to your needs may run on certain computers but not on others. At the very least, you should start with some idea of the most important things you want a computer to do and be sure the one you choose has programs and technology available that will do those things well.

Many computer programs are not really teaching programs, but rather utilities that allow the user to do something useful that will also be instructional. For example, some utilities allow the user to draw pictures, do things to the pictures (such as move them in a straight line, reflect them in a line, rotate them around a point, duplicate them in a different place, and so on). Other utilities will graph functions. There are spreadsheets that can be used to help analyze data. Some utilities allow the user to write all the axioms of a deductive system into the program, and then the computer will check any purported proof to see if it is valid. There are utilities that help students make and test conjectures. Programs are available to help students run simulations to see what would be likely to happen in thousands of trials of the same experiment. And many more.

Undoubtedly one of the most useful and powerful kinds of utilities is the word processor. Word processors can (and should) be used in mathematics classes (as well as English classes) to help students improve their exposition about mathematics. They can be used for such mundane tasks as checking for spelling and typographical errors and thus free student and teacher to concentrate on content and style. They also allow for much easier rewriting and thus encourage better final products.

In general, utilities of the sort described here, and the many others that help the user actively pursue some goal, are of much more value in a mathematics classroom than are most so-called teaching programs. The best teaching programs often have something similar to these utilities that allows the learners to be intellectually active and to follow their own goals.

The following example, like the calculator race, is designed to help students learn when it is appropriate to use what technology.

Suppose that you go to a fast food store and buy a hamburger for 99 cents, a milk shake for 75 cents, and a bag of french fries for 40 cents. How much will this cost you? What is the best way to figure it out?

With a small amount of mental mathematics skills, most people will note that 99 cents is 1 cent less than a dollar, and 40 cents can easily be broken into parts of 25 cents and 15 cents. Thus, the shake and fries together cost $1.15, and the hamburger will add 1 cent less than a dollar, so the total is $2.14.

This problem was easily done in the mind. Some people may have wanted pencil and paper or even a calculator, but they don't seem necessary.

Now, suppose you are entertaining friends, (or have an incredible appetite) and you order 7 hamburgers, 6 milk shakes, and 8 bags of french fries. Even though you may want to do some of the calculations in your head, you are likely to be a bit more comfortable using a calculator.

Next, suppose you own the store and want to find out how much income you should have had last week if you sold 7,349 hamburgers, 5,296 milk shakes, and 6,934 bags of french fries. You could use a calculator, but if you have a spreadsheet utility on your computer, and have programmed it to solve this problem each week, you are likely to find it much quicker and more reliable than other methods. Furthermore, if you would like to find out how much you *would* have made if you changed the price of hamburgers to $1.09, you can do this by simply pressing a few keys on your computer with its spreadsheet.

The point of all of this is that children should learn how to use technology as a tool to help them solve their problems. They should learn when it is appropriate to use which technology, and they should always expect to do the real thinking about when and how to use the technology and what the answers produced by machines mean in light of their understanding of the situation.

Programming is important for anybody who wishes to do any substantial amount of mathematics with a computer. On the other hand, administrators, teachers, students, and others must keep in mind that programming a computer is *not* mathematics. The ability to program a computer may help a student do mathematics, but should not be thought of as something that can replace the learning of mathematics any more than typing should be thought of as something to replace learning English (or whatever the language of the land is). Thus, a course in computer programming (however useful) should not be counted in lieu of a required course in mathematics, any more than a course in typing should be allowed to replace a required English course.

Advances in technology are a reason for changing the mathematics curriculum and our methods of teaching mathematics. These advances also provide tools that can be used to help change content and methods. Much of the material, both print and computer software, that has been written to accompany the introduction of technology into the school is of very low quality. Decision makers must approach such materials with caution; but that should not delay or prevent the introduction of technology into the schools either because of the confused notion that such technology will interfere with the children's "real education" or because it is difficult to introduce.

References and Further Resources

Bitter, G. (1989). *Executive Summary, ASU/SUPAI Calculator Project*. Arizona State University, Community Services Center at Papago Park, Tempe, AZ 85287–0908.

Hembre, R., and D.J. Dessart. (1986). "Effects of Hand-Held Calculators in Precollege Mathematics Education: A Meta Analysis." *Journal for Research in Mathematics Education* 17: 83–99.

Kenelly, J.W., ed. (1989). *The Use of Calculators in the Standardized Testing of Mathematics*. New York: The College Board and The Mathematical Association of America.

National Council of Teachers of Mathematics. (1989). *Curriculum and Evaluation Standards for School Mathematics*. Reston, Va.: NCTM.

Papert, S. (1980). *Mindstorms: Children, Computers, and Powerful Ideas*. New York: Basic Books.

Willoughby, S.S., C. Bereiter, P. Hilton, and J.H. Rubinstein. (1981, 1985, 1987). *Real Math, K–8 Textbook Series*. La Salle, Ill.: Open Court.

6

Connections

The fourth standard in each section of the National Council of Teachers of Mathematics' (NCTM 1989) *Curriculum and Evaluation Standards for School Mathematics* is "Mathematical Connections." Under this title, the NCTM discusses the need for children to see the connections within mathematics and among topics such as probability, geometry, arithmetic, and so on. The council also emphasizes the connections between mathematics and other school subjects and between mathematics and situations children may meet in the world outside of school. In this chapter, I consider not only these connections, but the connections within mathematics from grade to grade (sometimes called vertical articulation). None of this is new. All of it is important. For centuries, people have argued in favor of strengthening all of these connections.

The tendency to divide knowledge into little compartments and to teach and learn one compartment at a time is a natural but nevertheless pernicious tendency in all schooling, from kindergarten through graduate school. Interestingly, the tendency is less pronounced at both ends than it is in the middle.

In the early grades, because we expect teachers to be knowledgeable about all appropriate curriculum content, we have no qualms about having one person teach all subjects. The teacher may choose to divide the curriculum into fairly small compartments (reading, writing, arithmetic, music, art, history, and so on) and may never show the connections between these various subjects. At least the opportunity is there, however, and in the early grades there is some evidence that teachers occasionally take advantage of the opportunity.

At the other end of the educational spectrum, we prepare people to work at the cutting edge of intellectual creativity. If they are to make serious contributions to knowledge and to the welfare of the earth, they generally will have to integrate information and methods from several different fields. Thus we see doctoral degrees, as well as Nobel prizes, being awarded in areas such as biophysics, econometrics, and astrochemistry.

Much of the education that goes on between kindergarten and graduate school, however, is very neatly compartmentalized. Even within mathematics, there is compartmentalization. The precollege curriculum is divided into arithmetic, algebra, geometry, and so on. At the college level and at the beginning graduate school level, the situation gets worse; for example, topology (a part of geometry) may be divided into point-set topology, algebraic topology, and general topology.

The products of the education system would be better served if education were not so neatly compartmentalized. The connections between mathematics and the real world, between mathematics and the various sciences, between mathematics and the social studies, between mathematics and languages, and between mathematics and the arts should be constantly emphasized, not obscured. Beyond that, the connections between various parts of mathematics should be made clearer than they now are.

Connections with Other Subjects

Previous chapters have shown how mathematics can be derived from the real world of the learner and how mathematics can be applied to problems from everyday life and various other disciplines. Such connections should permeate all the teaching of mathematics.

Applications of mathematics to the physical sciences are well known and are commonly seen in schools, both in mathematics classes and in courses in the physical sciences. This is good, but should be even more common than is now the case.

Applications to the biological and other life sciences, and to other subjects, such as politics, history, fine arts, practical arts, and English, are less commonly studied either in the mathematics classroom or in courses in those other subjects. Applications abound in reality, though they may be sparse in classrooms. We have mentioned several, such as Haldane's (1985) use of geometry to analyze shape and size and other attributes of various species of animals, as well as the use of mathematics by politicians to gerrymander the congressional districts of states. The following examples of applications of mathematics to English, art, and music are mentioned simply to give some indication of the richness and abundance of such examples, even in areas not commonly thought to be mathematical.

• Statistical analyses have been made of word length and sentence length of various authors (notably Shakespeare and Bacon) to estimate the likelihood that material attributed to each was written by the other, or that the two might have been the same person. As of the present time, the best evidence is that Shakespeare wrote his own plays and Bacon wrote his own science books.

• Statistical analysis of frequency of letters in any document written in the English language can be used, and has been used, to break various codes. The same is possible in other languages; but, of course, the frequencies differ from language to language. Complex uses of mathematical number theory, involving the factoring of very large numbers into their prime factors, have gained considerable notoriety as ways to encode confidential messages and as ways to break those codes.

• Projective geometry is closely related to drawing with perspective. The assumption that parallel lines meet in an idealized point at infinity is basic to projective geometry as a point of perspective is standard in drawings with perspective.

• The "golden section" is the result of dividing a line of length q into two lengths, r and s, so that the ratio of q to r is the same as the ratio of r to s. A simple mathematical derivation involving the quadratic formula can be used to show that the ratio of r to s is equal to $(1 + \sqrt{5}) \div 2$, or about 1.618. This ratio (approximately) appears in architecture, art, various ratios of parts of the human body to each other, and in numerous other places that are thought to be aesthetically pleasing to the human eye.

• George David Birkhoff (1933), one of the leading mathematicians of the 20th century, wrote an entire book on the mathematics of aesthetics and on a mathematical approach to ethics. Although many people find some of his approaches and conclusions a bit "far out," such attempts are interesting and can be used to teach some fairly interesting mathematics as well as art and philosophy. More recently, Douglas Hofstadter (1979) wrote a book titled *Gödel, Escher and Bach*, which details many connections between mathematics, art, and music.

• The ancient Greeks knew that if the length of a musical string is divided in the ratio of 1:2, then the note of the shorter section is exactly one octave higher than the note of the longer and that musical notes could, in general, be described using mathematics.

• Any child learning to keep time in music will need to deal with fractions. Having studied and understood fractions will be of considerable help to such a child. Of course, studying music can provide good motivation for learning the mathematics.

Unfortunately, teachers of mathematics, as well as teachers of other subjects, are often ignorant of these many connections, and even when they aren't, they often find that the curriculum is too full without discussing such connections; or they believe that the discussion of such connections does not fit conveniently into the particular course they are teaching this year. Thus, children grow up remarkably oblivious to the many interesting and beautiful connections between mathematics and other branches of human thought and activity. Textbooks are often of no particular help in this matter, though there are some remarkable exceptions. *Mathematics, A Human Endeavor* (1970) by Harold Jacobs is certainly one such exception. An important criterion in choosing textbooks, and in choosing teachers, should be to encourage children to see and appreciate the connections between mathematics and the rest of the world.

If the subject of this book were physics, or English, or social studies, or something else, I would probably make the same statement, replacing "mathematics" with the other subject. A good liberal education does not consist solely of learning a great many things in isolation. A liberally educated person understands the connections between and within various branches of knowledge and is able to think and communicate rationally about them.

Connections within Mathematics

Mathematicians who are trying to solve problems, either within mathematics or involving applications of mathematics to other subjects, do not artificially limit themselves to one branch of mathematics, such as geometry, arithmetic, probability, or algebra. Nor do they even limit themselves to using only mathematics. In real life, people use whatever tools are available to solve their problems.

Why, then, do we teach as though crossing the artificial bounds between geometry and algebra, say, is somehow against the rules?

Pythagorean Theorem

The geometry teacher, for example, claims to be teaching a better understanding of proof by limiting students to geometric methods when proving the Pythagorean theorem. The standard proof seems like arcane drudgery to many students. Some of those students might find the following proof both more convincing and more elegant. Examine the picture of a square within a square (Figure 6.1) and the simple statement: BEHOLD!

FIGURE 6.1

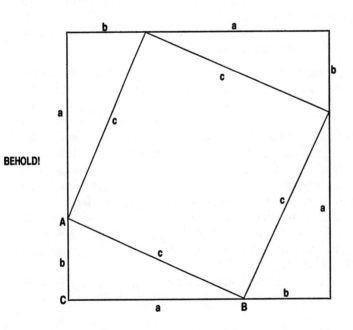

BEHOLD!

Only someone who is prepared to apply a bit of algebra to a geometry problem could figure out why this is a proof, and even then it may not be quite as obvious as the optimistic creator of the proof thought. For some students, however, this proof is likely more convincing, and easier to comprehend, than the standard proof or most of the alternate proofs that have been proposed over the years (there is an entire book of such different proofs of the Pythagorean theorem [Loomis 1968] proposed by people from all walks of life, including one President of the United States).

The details of the proof would go something like this: The area of the big square is $(a + b)^2$, the area of the small square is c^2, the area of each of the right triangles is $\frac{1}{2}(ab)$. So, setting the area of the big square equal to the sum of the areas of the four triangles and the small square, we have:

$$(a + b)^2 = 4 \times \frac{1}{2} ab + c^2$$
$$\text{or}$$
$$a^2 + 2ab + b^2 = 2ab + c^2$$

From which we conclude that $a^2 + b^2 = c^2$.

Of course, the construction of the original figure and the proof that various figures are, in fact, what we say they are, requires a bit of work; but the essential nature of the proof is immediately obvious to anyone who thinks at all seriously about it. That is an elegant argument. Yet, for many years, and still in some classrooms, that proof would have been declared inappropriate, unsuitable, or illegal.

If the goal of the geometry teacher is to teach about mathematical systems, and show how we can derive various theorems from a small set of axioms, and so on, a much smaller system than all of Euclidean geometry would be far more appropriate so that children can really see the connections. But even when mathematicians work within axiomatic systems, they usually are willing to accept some material from outside the system, such as arithmetic or the algebra of real numbers. If we are trying to teach geometry as an axiomatic system, why not start by assuming the algebra of real numbers, develop coordinate systems (with or without right angles), and use both algebraic and synthetic proofs?

Better yet, why not limit ourselves to a small portion of mathematics, such as incidence geometry or triangles, have the children conjecture and test their own theorems, and then see which can be derived from which others—thus creating their own mathematical system?

Regular Solids

Opportunities abound throughout school mathematics to use the interrelatedness of various mathematical topics, rather than to behave as though there were no connections. Let's consider a very simple example.

We are studying probability in the 7th grade. We have done many experiments to get an intuitive feeling for what happens under various circumstances. We have flipped coins and rolled cubes with numbers on their six sides (some people refer to such cubes as dice). We'd like to try some other objects that will generate certain numbers with equally likely expectations. We could try a spinner, but unbiased spinners are difficult to produce. Could we make dice with fewer than six sides or with more than six sides?

What kinds of conditions would such dice have to satisfy? A bit of thought usually convinces students that if all the faces were congruent regular polygons, there would be an equal probability of its landing on any one of its faces when rolled. We call such a figure a regular polyhedron. One can also imagine a die that is quite different from this that would still have equal probabilities of landing on any face—there would probably just not be as strong an intuitive reason to suppose that such a die would be "fair" (that is, have an equal chance of landing on any side).

Can anyone imagine a regular polyhedron other than a cube? Has anyone ever seen such a figure? Usually children will have seen several such figures and can imagine others. A triangular pyramid usually occurs to somebody.

Next, we can start considering how many regular solids there could possibly be. By considering how many faces could meet at a vertex, we can easily show that no more than five regular polyhedrons are possible (the sum of the angles could never be as great as 360 degrees, or you couldn't fold the pattern). Children can then try to create two-dimensional patterns for those regular solids and, by actually constructing them, prove that there are exactly five regular polyhedrons.

Looking for patterns between the numbers of faces, edges, and vertices of the solids ought to result in conjecturing a theorem about that relationship (sometimes known as Euler's formula), which would be written using algebra, though it is classified in the branch of mathematics known as topology. Concepts of duality may even be raised if students look for relationships between the various solids.

We have now used some algebra, quite a bit of arithmetic, solid and plane geometry, and many of our procedures for solving problems to investigate a problem that began with probability. The project, in this case, started with a problem in mathematics that was fairly closely related to problems in the real world, and our discussions would move quite comfortably from abstract considerations to very practical considerations (for example, where would you put the flaps to stick the solids together?).

Vertical Connections Among Grade Levels

The kinds of connections discussed so far are relatively easy to assess—and even fairly easy to institute. Connections among grade levels, or "vertical" connections, are much harder to evaluate and to make happen. Seldom do pupils have the same teacher from year to year: moreover, it is not uncommon for the teacher of a particular class to change during the academic year.

Teachers at one grade level (say, 2nd grade) seldom discuss with other grade-level teachers (say, 1st and 3rd grades) what has been or is going to be taught. Written records help if they are written properly and read carefully, but teachers generally do not have time and energy to perform either of these tasks in an optimal fashion.

Some people believe that curriculum guides alleviate the seriousness of the situation, but these guides seem to be honored more in the breach than in the observance. Textbooks and standardized tests are the principal influences on what actually gets taught in classrooms. This is not surprising, because textbooks offer material to actually help carry out the day-to-day teaching, and both teachers and entire schools (as well as children) are often judged by the results of standardized tests.

Textbook Adoption

Textbooks ought to be a major force to encourage a coherent program from grade to grade, building on past activities and preparing for future learning. However, many school systems adopt textbooks from different publishers for different grades, so that even if the textbooks have a coherent plan, that plan doesn't affect children in the classroom. In some cases, the school system, in the name of freedom, will adopt several different series and allow each individual classroom teacher to choose from the list. The results are even more chaotic than those from adopting different series for different grades, because pupils exposed to several different textbooks in the previous grade will all be in the same classroom.

Even when a school system adopts a textbook series from a single publisher for a long sequence of grades, say 7th grade through 12th grade, the authors may have changed from grade to grade in a way that makes it impossible to suppose there is any continuity.

Textbook adoptions made by single-grade committees or, at most, by committees that consider two or three grades, exacerbate the situation. Some years ago one of the large adoption states adopted a program from a major publisher for grades K–2 and for grades 6–8, but the committee for grades 3–5 found the program wanting. Such decisions—and the procedures that lead to them—discourage publishers from producing textbook series that are vertically integrated.

Functions

To make this abstract discussion about vertical connections more concrete, let's consider the topic of functions. The concept of functions is one of the most pervasive and important topics in mathematics. There have been numerous recommendations, at least since the 1890s, that this subject receive more and better attention in precollege schooling.

How can this topic be studied in a way that builds on the learner's previous understandings, relates the topic to other mathematics and other subjects, and is coordinated through the school years to prepare for what is to come and to build on what has gone before?

Kindergarten. The kindergarten class can get a large cardboard box from the grocery store and decorate the outside to make it look a little like a computer. Cut a small slot in the front to allow objects to be put into (and out of) the box. Then, while nobody else is looking, the teacher asks one child to get inside the box with appropriate materials (such as pencils, erasers, and other small objects common to a classroom) and a rule (such as "add two").

A few minutes later, the teacher suggests to the class that the box really is a "magic number machine," and if children wish to see how it works, they can volunteer to put something in to see what happens.

The first child approaches and puts a pencil in the machine. The machine rocks back and forth, bounces up and down a couple of times, may make some strange noises (depending on the talents and inclinations of the child within) and spits out three pencils. The class is usually impressed with this result (and, typically, some child looks around the class to see who is missing).

A second child puts in an eraser. The machine returns three erasers.

A third child decides to put in two pennies. What will come out? "Four pennies" and "six pennies" are both reasonable answers and should be commended. The child puts the two pennies in and out come four pennies.

What will happen if somebody puts in three ice cream sticks? What would happen if somebody put in eight crayons? What would happen if Manolita climbed into the machine? Why does nothing happen when zero dollars are put into the machine?

All of these are reasonable questions to ask, though only the first two are easy to answer. Even if children agree that Manolita wouldn't fit into the machine (at least not through the slot), they like to speculate about what would happen. Would we get three Manolitas back or one Manolita and two other children, or something else? As for the zero dollars, there usually is some general belief that you have to "wake the machine up" or stimulate it by putting something in before you can reasonably expect to get anything out. Such discussions are generally constructive, and they call attention to the difference between fantasy (or theory) and reality; but the main point is to encourage the children to try to predict future events on the basis of the patterns they have seen in past events.

At a later time, a child using a subtraction rule can be in the box. This makes part of the job easier; the child needs no materials since a smaller number of things will be returned than are put in, but subtraction can lead to certain difficulties. Suppose the rule is "subtract three" and only two things are put in. In one kindergarten class where this happened, the machine promptly announced "I want one more stick!"

Grade 1. In the 1st grade, essentially the same machine can be used, but in a slightly more abstract way. Instead of using objects, use small slips of paper with numbers written on them. The machine then crosses out the written number, writes a different number, and passes the slip back. So, if the rule is "subtract two," and "5" is written on the slip, the slip will come back with the "5" crossed off and a "3" written on the slip.

With numerals written on a piece of paper, there is now a possible answer to the question that arises when the rule is "subtract two" and "1" is put in. If children have been exposed to negative numbers (and it is entirely appropriate that they should have been), "–1" (negative one) is a possible answer, and "impossible" is also correct, depending on what kind of things we are pretending we are talking about. If the numbers are temperatures, bank balances, heights above sea level, or something else for which negative numbers are reasonable, "–1" would be correct; but if they are thought to represent objects such as sticks, pencils, or coins, then, of course, "impossible" would be correct.

Grade 2. By the 2nd grade, more complicated rules (possibly involving multiplication) and bigger numbers can be used. After several experiences, the activity can be taken to a higher level of abstraction by drawing a picture of the box and showing several pairs of inputs (numbers going in) and outputs (numbers coming out)(see Figure 6.2). Then, children can try to guess the function rule, or may be given a function rule and asked to supply missing outputs or inputs.

Figure 6.2

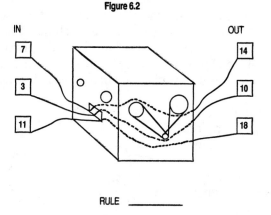

IN

OUT

| 7 | | 14 |

| 3 | | 10 |

| 11 | | 18 |

RULE _____

Grade 3. In the 3rd grade, the box can either be omitted altogether or used only briefly. Even the pictures of the box can disappear. They can be replaced by relatively abstract symbols involving arrows, and letters can be used to stand for numbers (see Figure 6.3). Then students can, for example, find the output when 5 is put into a machine that multiplies by 3, can determine what the input was when 15 comes out of a machine that subtracts 5, or can find the rule for a machine given several pairs of inputs and outputs.

Children having trouble with any of these concepts would, of course, be led back through as much of the physical, manipulative basis for the work as necessary. Students are now ready for composite machines in which the result from one machine is put into a second machine. So, for example, if 5 is put into a "times 3" machine and the result is then placed in a "plus 7" machine, students can figure out that 22 will come out. Or, if they know some number was put into a "times 3"

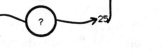

FIGURE 6.3

machine and the result was run through a "plus 4" machine producing 10, they can figure out that 2 must have been the original number (see Figure 6.4).

FIGURE 6.4

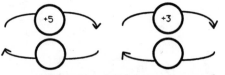

Third grade students quickly figure out that an efficient way to solve problems in which the output is known but the input is not known is to "run the machine backwards" or "reverse the machine." So, for example, if they know that 10 came out of a "plus 4" machine, they know that 6 must have gone in (10 − 4 = 6). Similarly, if they know that 6 came out of a "times 3" machine, then by running the machine backwards, they discover that 2 must have gone in.

This leads naturally to the idea of an inverse operation: when you run a "plus 5" machine backwards, it subtracts 5, or if you run a "divide by 3" machine backwards, it multiplies by 3 (see Figure 6.5).

FIGURE 6.5

The observant reader will notice that if a child has answered the question of what went into a "times 3" machine to give a result that went into a "plus 4" machine to produce 10, then the child has, in effect, solved the linear equation $3n + 4 = 10$.

In 1975 I field tested materials that followed essentially the program outlined here. One of the 3rd grade teachers had taught 9th grade algebra the previous year. One day after class, she cornered me and asked whether what the children were doing was not awfully similar to the solving of linear equations with which her 9th grade students had such serious difficulty the previous year. I admitted that there was a good deal of similarity and asked whether her 3rd graders were having any trouble with the material.

"No," she responded, "but my 9th graders had a lot of trouble with linear equations last year. I don't think you should do this in the 3rd grade."

If a proper foundation that relies appropriately on physical manipulatives and other aspects of the learner's reality is built over a period of years, young children can learn important concepts that might escape their older brothers and sisters who have not had the necessary preparation. The fact that people without appropriate preparation fail to learn something does not, of course, suggest that younger people, with the appropriate foundation, should not be taught the concept—especially if it is as important as the function concept.

Most elementary school teachers do not find the work described here with functions either difficult or distasteful. They see it as a fun, appropriate way to practice arithmetic, which it is. This calls attention to the fact that we are integrating algebra with arithmetic through this process, and showing how the algebra grows quite naturally out of arithmetic. It is not surprising that many elementary school teachers also see this work with functions as practice in recognizing patterns, since functions are, in fact, specific kinds of patterns.

Grade 4. In the 4th grade, children can continue practicing the use of function machines and arrow arithmetic with more complex rules; but they can also begin to integrate their work in algebra with geometry.

Children should begin learning the power of graphing early in their mathematical careers, and they can integrate it with algebra through the graphing of functions by 4th grade. In kindergarten and 1st grade, children can begin to learn about bar graphs by putting a counter with a hole through it on a peg for each time a certain event occurs (a car goes by on the street outside the window, for example). Several different kinds of events (a pedestrian walks by, a vehicle other than a car goes by, etc.) can be recorded at the same time on other pegs with similar counters. Then, a picture of the piles of counters becomes a bar graph. Keeping track of events with tally marks is a slightly more abstract approach to bar graphs.

Two-dimensional graphing can be modeled first with city streets and avenues, labeled 1, 2, 3, and so forth. To decide where the corner of Third and Fifth is, you must decide whether streets or avenues will be reported first (or give the entire name). Then locating points in the "city" is equivalent to locating points on a graph. Building on this background, and the graphing of such things as a child's height over time, students should be ready to graph linear functions and similar relations in the 4th and 5th grades.

Grade 5. If, in the 5th grade, the teacher introduces the standard notation for describing algebraic functions ($y = 7x - 5$, $y = x^2 + 2$, and so on), the students almost always complain that this simple notation has been withheld from them for so long and argue that they should have been taught with this "simple" notation from the beginning. Of course, the reason they find it so easy and natural is because they have gone through the relatively more cumbersome, but much more intuitive, procedures described here.

Grade 6. In the 6th grade, children can continue to graph more complicated functions, including periodic functions (average monthly temperatures for various cities throughout the year, for example), quadratics (the area of a square plotted against the length of a side, the area of a circle plotted against the radius, or $y = x^2 - 3x + 4$, for example) and can even find, or approximate, the values of x that make y equal 0, 10, or another number, for such functions by using graphs (and a calculator when the arithmetic becomes difficult). Thus, solving quadratic equations by appropriate means in the 6th grade is not at all beyond the powers of normal children, if appropriately integrated developmental material precedes this work.

Using graphing techniques with calculators and computers to solve equations also ought to make possible a substantial reduction in the emphasis on much of the exotic factoring that now occupies a major portion of first- and second-year algebra courses, thus saving time for more useful activities.

The grade levels reported here are not speculative. Tens of thousands of American children from all socioeconomic backgrounds—with all the usual variations in intellectual ability and with all the common physical, social, and personal problems—have successfully completed the work described here before the end of the 6th grade (along with equally challenging work in other aspects of mathematics). (See Dilworth and Warren 1980 and Herbert 1984 for evaluations and field testing reports.)

On the other hand, there is nothing sacred about these particular grade levels for these particular topics. To list each of the skills mentioned here as a requirement for that particular grade in a curriculum guide, and then reject any mathematics series that did not teach the given skills at that grade level, would be sheer madness. Rather, textbook committees and other decision makers should be looking for an approach to each important strand that is clearly integrated from grade to grade, that is based on the best information we know about how children learn mathematics, and that has been tested with real children in real classrooms. The general order of the topics described here demonstrates serious thought about human and mathematical development over a period of years. Evidence of such serious thought should be one of the most important criteria used in selecting textbooks.

Grades 7 and 8. In grades 7 and 8, students who have the kind of background described here can continue to study various kinds of functions, including exponential, quadratic, and other polynomial functions; periodic functions such as moon phases and tide heights; and other functions that originate in the real world around them. They should constantly connect this work with other parts of mathematics and with other subjects as well as with the work that has preceded this.

An example may help. Pretend that we are going into the turtle life insurance business. We must start by collecting some data. We find a sample of 10,000 newborn turtles and mark them so we can identify them later. We have reason to believe they are ordinary turtles who will live ordinary lives, and so we will be able to infer from them to the entire population of turtles for life expectancy. We have a limited amount of money to collect data, so we send a researcher out to count turtles only once every 20 years. The results appear in Figure 6.6.

This figure represents a function showing the number of turtles alive at age X. Given an age (such as 120 years) we can find how many turtles were alive then (7,800). Assuming this sample is representative of the population, we can now calculate the probability of a newborn turtle's living to the age of 120 (7,800 ÷ 10,000 = .78).

Suppose we want to calculate the probability of a 57-year-old turtle's living another 100 years. How do we do this? We first must decide how many turtles from our sample were alive at age 57. We don't know. We can't know because no one was around at the time and the turtles aren't talking. What do we do? We assume that turtles died at a constant rate between ages 40 and 60 (this is almost certainly a false assumption, but it's the best we can do). There were 9,000 turtles alive at age 40 and 8,900 at age 60, so 100 turtles died in 20 years. That's 5 per year. So, after 17 years, 85 turtles can be assumed to have died, leaving 8,915 live turtles.

FIGURE 6.6

Age X	Turtles surviving to age X
0	10,000
20	9,200
40	9,000
60	8,900
80	8,700
100	8,300
120	7,800
140	7,000
160	6,100
180	5,100
200	3,900
220	800
240	40
260	3

The next step is to determine how many turtles were alive at age 157. Between ages 140 and 160, 900 turtles died at an average rate of 45 turtles per year. That leaves 6,235 live turtles at age 157.

So, the probability of a 57-year-old turtle living another 100 years is about 6,235 ÷ 8,915 or .70.

We have just done the most difficult kind of interpolation for students in trigonometry and other courses—the kind in which one number is getting larger while the other is getting smaller. We have also seen both the reason that interpolations generally do not give exactly correct answers and the reason we tend to believe they are good estimates anyway. We've connected our study of functions (and algebra) with statistics and with probability. If we also had graphed the function, we would have shown a connection with geometry as well. And, of course, the work is based on an understanding of functions that was gained in preceding grade levels.

High School and College. Most of the study of mathematics in high school and college is directly or indirectly related to functions. Most of modern trigonometry is a study of circular functions. The consideration of periodic functions mentioned earlier should help prepare pupils for such study. Of course, we would start with some good concrete geometric work with right triangle trigonometry, and we would actually have students wrap a tape measure or marked string around a unit cardboard circle to derive the standard circular functions and the relations between them before going into the more abstract study of circular functions.

First- and second-year algebra and calculus could almost be described as not much more than the ever deeper and broader study of functions, their properties, and their applications. The study of all of these should build on a strong intuitive understanding of the function concept, built up over many years, and its importance both in mathematics and in the world around us.

There are, of course, many other ways functions should be studied and used in school mathematics, but those discussed here give an indication of the kinds of connections that ought to be made—both "vertically" and "horizontally"—when studying any topic.

Adopting a Connected Program

The purpose of this discussion of functions was to show how a topic in mathematics can be taught over a period of years. The foundation in the early years is built on the learner's reality, which is gradually expanded to involve serious study of very abstract mathematics. This study always relates back to the real world, other academic subjects, and other parts of mathematics. Similar development should occur for every major strand in school mathematics.

The typical classroom teacher is not prepared to make all the connections required for such an activity and should not be expected do so. On the other hand, textbook authors can quite reasonably be expected to do the thinking, planning, writing, field testing, and rewriting necessary to accomplish the kind of integration I have been discussing. Adoption committees ought to ask for some kind of evidence that this has been done. At the very least, they should ask who did the overall planning of the series and what kinds of considerations influenced that person or persons when making decisions about what topics would be studied at what levels, and how they would be approached.

Beyond that, however, textbook adoption committees should be divided into "vertical" subcommittees and "horizontal" subcommittees. For K–8, one subcommittee might look at the development of arithmetic operations, another might consider functions, another might examine the development of geometric concepts, and so on. For grades 9 through 12, functions could again be a strand to be studied; a second strand might be problem solving and applications; a third could study the presentation of geometric ideas in all four grades and how these ideas are integrated into the rest of the program.

A subcommittee should consider both the pedagogical development and the mathematical development. It should decide whether each is appropriate for children of the given age with the background provided by the textbook series. It should also decide whether the proposed activities prepare pupils for what is to follow.

If such adoption activities were carried on as a regular part of all major adoptions, the effects would be most salubrious for the quality of available mathematics textbook series.

References and Further Resources

Birkhoff, G.D. (1933). *Aesthetic Measure*. Cambridge: Harvard University Press.

Dilworth, R.P., and L.M. Warren. (1980). *An Independent Investigation of* Real Math: *The Field Testing in Learner Verification Studies*. Peru, Ill.: Open Court.

Haldane, J.B.S. (1985). *On Being the Right Size and Other Essays*. Oxford: Oxford University Press.

Herbert, M. (1984). *CSMP Final Evaluation Report*. Comprehensive School Mathematics Project. St. Louis: McCREL (Midcontinent Regional Educational Laboratory).

Hofstadter, D.R. (1979). *Gödel, Escher, and Bach: An Eternal Golden Braid*. New York: Basic Books.

Jacobs, H.R. (1970). *Mathematics, a Human Endeavor: A Textbook for Those Who Think They Don't Like the Subject*. San Francisco: W.H. Freeman.

Loomis, E.S. (1968). *The Pythagorean Proposition*. (Reprinted). *Classics in Mathematics*, vol. I. Originally published in 1927 with second edition in 1940. Reston, Va.: National Council of Teachers of Mathematics.

National Council of Teachers of Mathematics. (1989). *Curriculum and Evaluation Standards for School Mathematics*. Reston, Va.: NCTM.

Willoughby, S.S., C. Bereiter, P. Hilton, and J.H. Rubinstein. (1981, 1985, 1987). *Real Math, K–8 Textbook Series*. La Salle, Ill.: Open Court.

7

Fostering Change

In this book, I try to present arguments supporting the need for change and to indicate what kinds of changes are needed, occasionally even suggesting specific actions that can be taken to ensure those changes come about. Constructive change, of course, is very difficult to achieve; but there are various ways educational leaders can encourage positive change.

Metamorphobia

A characteristic that appears to be present to a greater or lesser degree in almost all human beings is a fear of change or an avoidance of change. I call this characteristic "metamorphobia." Two basic principles seem to govern the behavior of most people in this regard:

1. People try to create or acquire the simplest possible rules to explain situations and to govern their behavior.

2. People try to keep the rules that appear to have served them or others well in the past, even in the face of new evidence that those rules are not optimal.

Let me give some simple examples of what I mean. First, in early infancy we somehow identify a set of very different-looking perceptual images as a single thing that we ultimately call "mother." Much thought is saved by not treating each instance as a separate and unknown being. By doing this we have created a simple, and simplifying, rule.

Second, a young child naturally concludes that, in general, of two sets of discrete, identical objects, the one that covers the greater area has the greater number of objects. This conclusion is generally true, barring the intervention of some malevolent outside force (such as an adult psychologist). Even when the conclusion fails to be true, the child has neither the intellectual equipment nor the inclination to recognize the falsity of the proposition. For the normal preschool child, the conclusion relating area and number is easy to make and appears to work. Before Piaget, probably no more than one child in a million was ever embarrassed in the slightest wnen applying this principle. Since the principle seems to work, the child is loath to give it up.

Similar resistance occurs whenever people of any age are faced with information or theories that might cause them to question the simple principles they have previously acquired and used successfully. We see an analogous phenomenon in adult human beings who are presumed to be at the pinnacle of their intellectual professions. For example, the medical profession rejected antiseptic procedures and the whole germ theory of disease until long after the evidence was overwhelming.

Human beings at all levels follow the two simple metamorphobia principles: (1) create simple rules, and (2) try to retain previously known rules.

There are two obvious reasons that people behave this way. First, as individuals, we conserve energy, reduce the need for substantial thought, and avoid potentially dangerous situations. Second, as a society, we conserve the knowledge of the past (even when it happens to be false knowledge). The ability to transfer substantial quantities of knowledge and experience from one generation to the next is one of the main characteristics that sets the human race apart from other animals. It gives us the ability to build on the past rather than starting over with each new generation. This ability should be cherished and preserved.

These rational reasons that metamorphobia ought to exist do not, of course, prove that it does. Can it be tested? Perhaps by observation. Perhaps even by making a prediction or two about how people will behave in certain situations. Try to do this with your own behavior as well as with other people's (even though you may incur some difficulty remaining totally objective). Do you have a tendency to believe, for instance, that all people with some characteristic X (for example, having red hair or being physically attractive) will also have characteristic Y (for example, lose their tempers easily or be honest and nice) simply because you have previously seen two or three instances of this confluence of characteristics? Do you change your mind easily when you observe counter examples, or do you assume these are "the exceptions that prove the rule"?

Most people seem to resolve the conflict between "known" general theories and an easy acceptance of new ideas and facts in favor of the former. Such people are characterized as knowledgeable, mature, conservative, close-minded, reactionary, or bigoted, depending on who is doing the characterizing.

Some among us tend more often to resolve such conflicts in favor of the new. Such people are characterized as creative, young at heart, liberal, immature, flighty, or irresponsible, again depending on who is doing the characterizing.

I believe that the most important contributions to making the world a better place to live never fall completely within either category. In general, the people who really help the world become a better place tend to accept new ideas readily but at the same time to be very aware of the knowledge of the past, as well as the knowledge that others are producing at the time. They remain ready to accept or create new ideas when the evidence supports them, but they retain a healthy respect for past knowledge.

Historically, the educational system has been charged with the duty of passing on the knowledge of the past. This is as it should be, since knowledge of the past is required if we are to improve on it with any efficiency or even live comfortably in the present. Language, social institutions, mathematics, science, and other kinds of knowledge are all passed on from generation to generation, at least partially, through schooling. Although passing information on to the next generation is necessary and desirable, there are obvious difficulties with any system that only presents past information. First, there could be no progress if the human race were limited to only the knowledge of the past. Second, this knowledge is never transferred perfectly (witness the game of "telephone"). There is always some loss and some distortion. So, while valuing those who bring us knowledge from the past, we must also cherish those who bring doubt and creativity to bear on such knowledge.

Formal schooling is one of the prime mechanisms society uses to instill the wisdom of the past in each new generation so that the human race can conserve thought and build on what has gone before rather than recreating, in each generation, what our ancestors knew. Given this important mission of the schools, we would expect them to be conservative and authoritarian. They are.

Teachers, textbook publishers, creators of standardized tests, parents, pupils, and the public at large all have a vested interest in the status quo. Life is easier and less threatening when we continue doing things the way we, and our predecessors, have always done them. How then can we expect the schools to change so as to prepare children for the challenges they will face in the 21st century?

I have no simple rule that educational leaders can use to govern their behavior to help themselves and others change mathematics education. I have, instead, only a few suggested actions that may be of some help in facilitating the sorely needed changes.

The Role of Teachers in Change

Other than the pupils themselves, teachers are the most important part of the educational process. The popular press and various political pundits have had a field day telling us what a bad job our teachers are doing. Low grade-point averages and low standardized test scores for people hoping to be teachers are cited to prove that teachers aren't good enough. It has been widely reported that U.S. teachers teach only 180 days a year and teach only 25 to 30 students in a class. These facts have been presented as apparent indications of how lazy and overpaid our teachers are, and also as reasons our schools achieve less at greater cost than schools in certain other countries, such as Japan.

But the reports seldom mention that more and better qualified people would choose to enter the teaching profession if conditions and compensation were better. Nor do they mention the fact that Japanese teachers teach only 16 hours in a six-day week while our teachers teach an average of 25 hours in a five-day week. U.S. teachers actually teach more hours in a year than Japanese teachers do, even though Japanese teachers are said to teach (and be paid for!) a full year.

Seldom do the reports call attention to the fact that the reason Japanese teachers are able to handle classes of 45 students is that there is strong family and societal support for education in Japan, and discipline problems as we know them simply don't occur in Japan.

Two surveys on discipline in U.S. schools, one taken in the 1940s and the other in the 1980s, suggest the difficulty of teaching school at present. The most serious discipline problems in the 1940s were talking, chewing gum, making noise, running in the halls, getting out of turn in line, wearing improper clothes, and not putting paper in wastebaskets. The corresponding list for the 1980s included drug abuse, alcohol abuse, pregnancy, suicide, rape, robbery, assault, arson, gang warfare, and venereal disease (Will 1987).

It has been said that anyone who is really qualified to teach mathematics and chooses to do so, knowing what conditions are like in the schools and knowing what the other options are, is truly committed—or truly ought to be. Given the conditions of teaching and lack of respect for education, there is an amazingly large number of truly qualified—truly committed—teachers of mathematics in the United States today. As a society, we should be grateful to them and should try to improve the conditions under which they labor, both to help them do their jobs better and to attract more of these excellent mathematics teachers for the future.

Another fact that is widely ignored by those who view our school system with alarm is that much of the money they claim to be "throwing at the problem of education" simply is not spent on traditional education. Much of that money is spent on busing; special classes for the physically and mentally handicapped; police-type protection for teachers and students; impressive physical plants maintained by janitors who are paid more than the teachers or other instructional staff; special courses about drugs, AIDS, dental hygiene, driving an automobile, and nutrition; and other activities that were not traditionally thought to be part of educating children in this country, and are not included in the education budgets of most other countries.

However worthy you may believe these many activities to be, it is simply not rational to pretend that the money spent on them is likely to improve the quality of education in mathematics, history, English, music, or science. The recent report of the Economic Policy Institute (1990), ranking the United States 14th out of 16 industrialized countries in expenditures for education, has generated considerable debate. But, even if we are closer to the middle of the pack than suggested by that study, we still spend far less on what has traditionally been called education than do the other industrialized nations.

Other than behaving as concerned, voting citizens and bringing pressure on elected representatives at all levels of government, there is little we as educators can do to correct the many problems with U.S. education. Even within this unfortunate context, there are many things we can do to help teachers do a better job. We can choose candidates who are likely to be effective teachers, help them to become and remain high-level professionals, and support their positive classroom activities through visits and conferences.

Choosing Teachers

In recent years, and still today in many situations, a discussion of "choosing mathematics teachers" would seem strange. Anyone who is certified and who applies for a position would automatically be offered the job. When nobody who is certified applies, anyone who can get temporary or provisional certification is hired. Failing that, a person who is certified to teach some other subject and who is willing to teach mathematics, even without the necessary background, would be impressed into service.

Whether there is a plethora or dearth of candidates for a teaching position, there are certain criteria that ought to be considered in determining whether the candidate is qualified. These include knowledge of content, ability to communicate about that content, knowledge of the people to be taught (e.g., a knowledge of psychology and sociology), general understanding of the place of education in society over the years (history and philosophy of education), a good general liberal education that allows the candidate to understand and discuss the connections between mathematics and many other areas of human thought, a personality that is attractive to both children and colleagues, and a substantial amount of energy and commitment so that the candidate will be able and willing to put all these talents together to do an excellent job of teaching.

Occasionally people raise a false dichotomy between a liberal education and professional education and seem to suggest one or the other is superfluous. Indeed, some states, and even some branches of the federal government, have suggested that anyone who is qualified to teach one subject is therefore qualified to teach any subject. This is analogous to suggesting that dermatologists and orthopedic surgeons ought to be freely interchangeable.

Others have suggested that if teachers just know the content, all professional education is useless; or, at most, all one needs is a couple of courses in the summer and maybe a supervised internship to become a qualified teacher. This is equivalent to suggesting that anyone with a degree in anatomy should be allowed to practice medicine, perhaps after taking some summer courses in medical practice and serving a brief internship.

The medical analogies are not meant to be facetious. Education is at least as difficult as medicine, and is more important to society (if not to the individual). If we don't take ourselves seriously, nobody else will. There have been instances of people practicing medicine without having attended medical school (sometimes with considerable success over a long period of time), just as there have been instances of people teaching successfully without the generally expected background. That does not mean that either practice is desirable on a global basis.

How a particular person acquires the requisite knowledge is another matter. If a potential teacher can study mathematics without attending school and learn as much as someone who has, there is no real need to expose the individual to the mathematics faculty of a college or university. The same goes for professional education courses. Evaluating such individual learning to see if it is really adequate would be difficult, but should be possible in both cases. The same should be true of someone who wishes to become a physician by nonstandard procedures—if the prospective physician or teacher can demonstrate all the knowledge and skill typically required, the technicality of whether a particular course or degree appears on a transcript should not be pertinent.

Several organizations have published standards on qualifications of mathematics teachers. The Mathematical Association of America (Committee on the Mathematical Eduction of Teachers 1988) and the National Council of Teachers of Mathematics (NCTM 1989b) are two such professional organizations. Several other groups are working on similar standards or have already published them. Certainly anyone who is about to hire a teacher of mathematics ought to refer to such documents.

Elementary school teachers usually are teachers of mathematics, and the organizations mentioned here have also made specific reference to the kinds of mathematics and professional courses they should have. Of course, similar consideration should be given to other subjects elementary school teachers will teach, such as reading, science, music, and art.

Good teachers who are already on the faculty should be involved in the hiring of new teachers. They will have insights different from those of a supervisor or administrator. Beyond that, if other teachers in the school have a voice in choosing a new teacher, they are likely to feel they have a stake in that teacher's success and are more likely to provide counsel and support for the teacher when it is needed.

If the criteria listed here and the criteria of various professional groups are used, rather than the much more limited, technical, legal minimum requirements imposed by the state, there is a high probability that no one who is qualified for the job will be found. Whenever that happens, the school leaders owe it to the community and to education to make the fact public. If the citizens of a community are told on a fairly regular basis that no one who is really qualified can be found who is willing to teach in their community, they may take action to improve the conditions of servitude for teachers.

Continuing Professional Growth of Teachers

Even if the best potential teachers are hired, they must continue their professional development to live up to their potential. To advocate education for others but refuse to continue one's own education is hypocritical. To be an honest supporter of education for others, a teacher must continue to learn.

A teacher's continuing education should be multifaceted. Every person whose primary responsibility is teaching mathematics should belong to appropriate professional mathematics teachers' groups. On a national level, the appropriate organization for pre-college teachers is the NCTM. The NCTM has local and state affiliates in every state, which serious teachers of mathematics should also join. Some high school teachers may also wish to consider membership in the Mathematical Association of America and the American Mathematical Association of Two-Year Colleges. Teachers should read the appropriate journals of the various professional organizations and should attend professional meetings of those organizations. Teachers, supervisors, and others often respond to such a suggestion with "Yes, but how are we going to find time and money for this given the other pressures of being a teacher?" The answer is not trivial, but often a simple reordering of priorities will provide an answer.

In one school system in the southeastern United States, a teacher who had been invited to speak at a regional NCTM convention was refused permission to go because she was needed in the classroom. Approximately one month later, the same teacher was required to leave her class to go to the state basketball tournament to check coats. The priorities of that school system were in dire need of reexamination (Willoughby 1984). In fact, after this story was publicized, priorities in this system were reexamined.

As a standard practice, part of the school budget should be set aside to help provide continuing education for all teachers.

Besides reading professional journals and attending professional meetings, a teacher can attend classes in colleges and universities (with or without credit), can work with other teachers in the school system in seminars or more formal classes to continue learning about mathematics and pedagogy, and should spend some time each year observing other teachers and being observed by others.

Reciprocal observations are almost always good for both teachers involved. The teacher who is being observed prepares an especially good lesson, which benefits the students of that class and can be reused in the future. The teacher who is observing sees an excellent lesson and can use ideas from it in future lessons. Even if the lesson is not exceptionally good, the visiting teacher may notice things to avoid doing.

Again, the question of how to provide time for teachers to visit each other is a serious one. At the risk of alienating some of my readers, I suggest that the answer is obvious. Principals and supervisors should, on a regular basis, substitute for teachers to allow those teachers to make such visitations.

The goal of schools is to educate. The main function of supervisors and principals is to improve that education. Unless they occasionally teach in the classroom, they will lose touch with reality and not be able to help teachers as much as they should. (By the same logic, of course, I believe that any teacher of methods courses or writer of textbooks should regularly teach at the appropriate level.) The word *principal* is not really a noun at all; it is an adjective. In schools, "principal" was originally part of the phrase "principal teacher." If supervisors and principals are really our best teachers (and they should be), then secretaries, clerks, and others should be taking care of the paperwork, administrative details, and public relations; and the educational leaders should be involved in the classroom.

Helping in the Classroom

Whenever possible, visits to a teacher's classroom should be for the purpose of helping the teacher do a better job, not for "evaluation" in the pejorative sense. Ordinarily, a good supervisory visit will be announced to the teacher in advance, and the teacher will describe the plans for that period. If a good textbook has been adopted, the supervisor should have a copy of the teacher's guide and should become familiar with the goals and procedures set forth for that lesson.

Recently, a staff member of a development project visited a field-test school. One of the participating teachers complained that certain preliminary activities should have preceded the first lesson. The staff member asked if the teacher had not read the teacher's guide for lesson 1, since it described all the teacher's activities and more. The teacher responded: "I never read instructions." Other teachers in the school concurred, saying they also never read instructions.

Society has serious problems when even teachers rebel at the thought of learning through the printed word because of the strain on their intellect or energy. At the very least, a visiting supervisor should see clear evidence that the teacher has read the teacher's guide. Virtually every teacher's guide ever published has some good suggestions for improving teaching. If the textbooks have been chosen carefully, there should be many such good suggestions for introducing and presenting concepts; for practicing skills in constructive, enjoyable ways; for alternate teaching strategies; for daily and longer term periodic informal assessment; and for remedial activities that can be carried out as soon as trouble is indicated.

If the teacher deviates substantially from the teacher's guide, the kind and quality of those deviations should be noted. For example, if the guide suggests doing some mental arithmetic, playing a game, working in groups on a project, and other such positive activities, and the teacher actually spends the entire period talking and having pupils do a page or two of written exercises, there should be some serious questions asked as to why the teacher chose to deviate in this way.

If, on the other hand, the deviations are in the opposite direction, with the teacher adding exciting activities that were not suggested in the teacher's guide, the teacher should probably be commended and perhaps asked to share some of those activities with colleagues so they, too, can "deviate" in this constructive manner.

Almost never should the teacher spend a long time talking "at" the class. It has been said of American education that people learn by doing but teachers teach by talking. Though some talking is appropriate, most teachers have a tendency to talk too much and allow too little time for involved participation. Mathematics is not a spectator sport. Teachers must endeavor to overcome the "sage on the stage" syndrome.

The teacher should be well organized and have prepared any necessary materials for the class. Beyond that, the children should be well organized. If there is written work to be done, they should have paper and sharpened pencils (without lining up at the sharpener after the assignment has been given). If a game is to be played or a project pursued, there should be an efficient method of distributing the necessary materials. If mental mathematics is being practiced, there should be some means by which the teacher can be sure everybody is participating and can also tell what mistakes are being made by which children. At the very least, all students in the class should know what they are supposed to be doing at all times. The teacher should also know both what individuals are supposed to be doing and what they *are* doing.

In a class in which pupils do many things other than listen and write, discipline is likely to be more of a problem than in more traditional classrooms; but good discipline is even more necessary because of the need to allow children to pursue their goals (play a game, solve a puzzle, complete a project, etc.) without undue interference from others. The beginning teachers with the greatest promise often have the most serious discipline problems precisely because they are trying to encourage intellectual freedom and have a hard time distinguishing in their minds and the minds of the pupils between intellectual freedom and social irresponsibility. An experienced supervisor can be a great help here. Such a teacher should be encouraged to solve the discipline problem quickly, even if some intellectual freedom must be temporarily abridged, lest all freedom be eliminated by the irresponsible acts of a few.

A spirit of cooperation, sportsmanship, inquiry, flexibility, and enjoyment should permeate the classroom. Thus, the teacher and students should listen to each other's views and respect different, but reasonable, answers. When playing games, players can reasonably be expected to help their opponents do better. When investigating an interesting mathematical question, students should show reasonable signs of interest and even excitement, as well as respect for others' opinions.

There should be a reasonable balance between "getting on with the work" and allowing time for deeper consideration of interesting problems. If the problems are really interesting, and the students are truly interested, there is very little excuse for rushing on to the next topic. On the other hand, once the interest of most class members has flagged or there seems little hope of learning anything new from continued consideration of a problem, moving on should be encouraged—perhaps with the suggestion that individuals pursue the matter further outside of class, in cooperation with other class members, by themselves, or with the help of other people.

Textbook Selection and Use

Studies about the influence of textbooks on the mathematics curriculum invariably conclude that the most important single factor in determining what content children are exposed to in the classroom is the textbook. In almost all mathematics classrooms in the United States, virtually nothing is taught that is not in the textbook. Thus, choosing textbooks and other curricular material is one of the most important activities in the educational process.

Despite the importance of textbook selection, texts are often chosen in a remarkably cavalier manner. During my first year of teaching in a junior high school, I was told I was on the mathematics textbook selection committee and should report to a particular room at 2:30 that afternoon (the time the last class ended). Two other teachers and I appeared in the room and began looking at the 10 to 12 series of textbooks that had been submitted. I observed that my more learned and experienced colleagues each chose a favorite topic, checked each series to see whether the book had done it "right," and quickly assigned each series to the reject or acceptable category. Several series were also rejected because they didn't appear to be attractive enough to somebody. By 3:30 we found ourselves with three series that had not been rejected. We voted and left. The series we adopted was the same series the school had used for the previous five years.

There are better ways to select textbooks. The NCTM (1984) has published a list of professional standards for selection and implementation of instructional materials that is well worth reading by anybody involved with textbook adoption. The first standard is:

The entire process of textbook selection should be led by teachers and supervisors with expertise and responsibility in mathematics education.

Except in most unusual circumstances the recommendations of the committee must be followed.

Although this recommendation may seem so obvious as to not be necessary, the NCTM committee that wrote these standards had been surprised by results of their survey that indicated that often textbooks are adopted by majority vote of the uninformed (teachers who in many cases had not even seen some of the books they were voting on) or by a political process often more dependent on which sales representative had made friends with the largest number of school board members than on any professional considerations.

Another NCTM recommendation is that criteria to be used in selecting books be published in advance. A short list of criteria is of far more use in such selection activities than a detailed, telephone-book-size, curriculum guide. There should be no attempt to specify exactly what is to be taught (or is not to be taught) at each level, nor exactly how each topic is to be taught. The general flavor and methods, the need for vertical and horizontal integration, appropriate level of expectations, and similar matters can quite reasonably be specified, however.

NCTM strongly recommends that those responsible for writing the criteria first spend a substantial amount of time studying various recommendations of professional organizations and others, and making decisions as to which of the goals specified by national groups they believe to be most important and appropriate for the local community. The organization recommends consulting current research and evaluating the presently used program to see where it is particularly strong and where improvement would be desirable.

Two additional comments made by NCTM are interesting because so many textbook adoptions seem to violate them. "Difficulties in implementation should be considered only if they seem insurmountable" and "Copyright dates and formula-determined reading levels are generally inappropriate as selection criteria."

NCTM recognizes that it will always be more difficult to implement a program that requires change than one that does not. Thus, if an adoption committee starts with the belief that any program that is a little hard to implement should not be adopted, there will never be substantial change. However, the standards suggest that substantial inservice support be provided whenever a new program is instituted. Such inservice support should precede the actual adoption and should continue well into the first year of the adoption. For really substantial changes, inservice support may reasonably continue for several years.

The point about copyright dates and formula-determined reading levels is particularly important in mathematics: a book with an old copyright date may be more up-to-date than a book with a more recent date. Reading-level formulas have been instrumental in destroying the quality of the language used in "word problems" and other exposition and should not be used.

Adoption committees ought to look at all available materials even if there is no local sales representative for them. Some excellent materials have been developed by nonprofit groups and by very small publishers. A serious effort should be made to locate all such materials that have any chance of being adopted. If the distributors are so small that they will not be able to supply consultants and other implementation support, efforts should be made to find out what sort of support would be available from other sources. Often there are good independent consultants who can fill such a gap, or hiring an additional member of the supervisory staff may be necessary and appropriate.

After the committee has reduced the number of serious candidates to two or three programs, the remaining programs should be piloted in local classrooms, if possible, to find out how they work in practice. Then, committee members should visit pilot classrooms; should listen carefully to both positive and negative comments from teachers, pupils, parents, and others associated with the pilot; and should make their final decision on the basis of all available information. If, for some reason, a pilot test is impossible, the committee (or some subset of the committee) should visit other school systems that are using the materials and see how the program is working there.

Occasionally people object to the suggestion of pilots or visits on the basis that new programs could not be considered since they would not have been used in the year preceding adoption. That, of course, is not true if the programs are being developed with any kind of decent respect for how they will work in the classroom, since proper development would certainly require serious field testing for at least a year before publication, and for several years if a series is involved and there is any attempt at vertical integration.

In the previous section, the point was made that teachers should continue their education throughout their teaching careers. In addition, whenever a truly new program is adopted, there should be strong inservice preparation and support. The contract to buy the books should include an agreement by the publisher to provide appropriate inservice education for teachers, supervisors, and others associated with the adoption. Usually a day or two for most teachers will be as much as can reasonably be expected, but it is not unreasonable to suggest that certain key teachers be given substantially more preparation so they can become resource teachers for others in their school.

Beyond this preparatory inservice work, there ought to be some systematic follow-up a few weeks into the program—a refresher course in which teachers can ask pointed questions about things that don't seem to be going quite right. Ideally, the publisher should have someone available at the other end of a telephone line (preferably an "800" number so the school or teacher doesn't have to pay for the call) who can answer most questions or can get the answer to questions and respond. But beyond that, a consultant should appear in the school regularly for the first year, and occasionally even in later years.

The school system itself should create a procedure allowing teachers who have trouble with a particular lesson to discuss it with others—to see whether they too had the same trouble and what they can do to improve the situation. When teachers join the system, there should be an automatic procedure for ensuring that an experienced teacher who is familiar and comfortable with the textbooks is assigned to acquaint the new teacher with the material and help smooth out the rough spots.

Supervisors, principals, and other leaders should also have some inservice preparation to become familiar with the goals and procedures of the new program and to be able to help teachers implement it effectively.

After a new textbook series has been in use for about a year, serious evaluation of the success of the program should be undertaken. If the evaluation points to weaknesses, appropriate modifications ought to be considered. Is more teacher preparation required? Would certain supplementary materials help? Should other modifications be made? As a last resort, should a different set of books be considered?

This last question is usually not considered seriously because of the expense involved, but changing books a year early often turns out not to be as expensive as some people think. For grades K–2, most student books are paperback "disposable" items that must be replaced every year anyway. For the other grades, if the cost of the material is prorated over five or six years, the actual cost turns out not to be terribly great. If a truly new and more challenging program is to be adopted in a future year, adopting K–3 early may help prepare pupils for the different and greater demands to be expected in grade 4 and above.

This discussion of cost brings us to a really important point about textbooks. The process suggested here seems to require a great deal of time, effort, and money. It does. The committee must be given time and substantial support to carry out its duties. However, the overall cost of textbooks in the United States is less than 1 percent of the total school budget (and also less than the nation spends on Nintendo). The influence of textbooks on the education of children is out of all proportion to their cost. There is no excuse for not taking the adoption of textbooks far more seriously than is presently the case in most schools. Beyond that, as the NCTM stated, because of textbooks' small portion of the school budget, cost should not be a determining factor in selecting textbooks. "Potential financial savings do not justify the selection of less desirable materials" (NCTM 1984).

One more point ought to be considered. If a textbook series is adopted that is considerably better than the previous series, children coming through the new series will be expected to know a great deal more and be able to do many more things than children who used the old series. Under those circumstances, with a K–8 adoption, for example, the upper grade teachers and students will have substantial difficulty because of lack of preparation. Parents, pupils, teachers, and others ought to be warned about this in advance. Expectations for the first year ought to be somewhat lower than for subsequent years because a lot of time is likely to be spent remediating the weaknesses of the old program. Obviously, failure of many classes to finish the book should not be seen as a weakness of the new program (though it might be seen as a weakness of the old one).

Assessment of Pupil Progress

Various methods of assessment should be used throughout the educational process in mathematics. Response exercises of the sort described earlier should give teachers immediate feedback. Observation of children playing games, doing activities together, responding to oral story problems, and discussing other special situations designed to foster thinking should show different aspects of each child's achievement. Written work done individually by the children provides still a different opportunity for assessment.

The primary purpose of all this evaluation should be to give the teacher a clear picture of the individual differences that exist within the class and to provide an opportunity to help children in need of help and challenge children in need of challenging.

"Individualization" has been in and out of favor at various times in the recent history of education. Nobody can object to the usual professed goals of individualization (to meet the needs of each individual child, or something similar), but solitary confinement with a ditto sheet, or opportunity to go "at your own speed," which for some children turns out to be zero, is not the best way to provide each child with the best education possible. The teacher should be as aware as possible of each child's progress. Individual progress records should be kept of children's accomplishments and needs; and as the needs turn into accomplishments, that progress should be recorded.

When a teacher discovers that a particular child is unable to do something that is necessary for further progress, the teacher should evaluate the situation further to decide whether the child does not understand or simply needs extra practice to become proficient in the skill. If understanding is deficient, the teacher will have to spend extra time with the child until the deficiency is remedied. When the problem requires nothing more than extra practice, the teacher can usually prescribe appropriate practice in the form of a game or some other activity that can easily be supervised by parents or others, or possibly accomplished without supervision.

Children ought not to be sorted out by somebody's idea of ability at an early age. There will certainly be times when individuals fail to understand certain concepts and will be in need of considerable extra help. It is also likely that certain individuals will need help far more often than others. That's the way the world is. Good textbooks should help good teachers provide that extra help.

Rather than giving up on those children who have serious difficulties or are different from typical successful mathematics students in other ways, we must do a better job of helping them. We must be alert to differences among children whenever they occur. We must have high expectations for all children, and we must give them the necessary help to live up to those high expectations. All children can and should learn mathematics.

Evaluation should be an ongoing, important part of education. However, the pathological fascination with grades and public assessments of children, teachers, and schools that permeates our society has no place in good education. Almost nothing useful is gained from such activities; and a great deal of damage can be caused when teachers teach for, and students study for, a less-than-perfect test. Virtually all schools would be better off with fewer standardized tests—and with less public dissemination of test results.

At the very least, schools should try to avoid any activity that would encourage teachers to teach for such tests and children to study for them, since that is not only a waste of good educational time, but distorts the function of the test. Standardized tests are designed to sample a small portion of the learning the school is expected to deliver. If nobody "preps" for such tests, they can be a fairly good indicator of the quality of the rest of the education. But that sampling procedure is completely destroyed by specific test preparation. When "test data" are published in the local newspapers or otherwise used to evaluate children, teachers, and schools, the motivation to teach to the test rather than for the education of the children becomes almost irresistible.

Some states and localities, and even some test makers, have given up trying to get people to use tests more intelligently and now hope to improve education by improving the tests. Some of the tests from this new generation are significantly better than the old ones. These new tests certainly should be used if such tests are going to be used—and abused—as they have been in the past. But we'd be better off if tests were used in a more intelligent way.

Every mathematics course ends with a final examination. So as not to disappoint the reader, I provide the following final examination and solution key. Please try the test before looking at the solutions.

Test

Preliminary Information

A.   I have ten fingers altogether and, for this test, assume everybody else does. That is, I do count thumbs as fingers, so I have a total of exactly ten fingers, not eight.

B.   Chris (a 2nd grade teacher) and Pat (a mathematician) are married and have a four-year-old daughter, Wendy. One day Chris came home from school and found Pat teaching Wendy the addition facts, whereupon Chris said: "You really shouldn't do that until we are sure she conserves number—you may be doing more harm than good."

Questions

1.   Write a four-letter word (beginning with "J") for an amusing anecdote.

2.   Write a four-letter word (starting with "Y") that is sometimes used to refer to a pair of oxen, and is sometimes applied to a wooden frame holding the two oxen together. (Spelling counts!)

3.   What do we call the white of an egg? (Spelling still counts!)

4.   How many fingers are there on ten hands?

5.   What is Wendy's mother's name? (Do not look back at the story.)

6.   What do we call the yellow of an egg? (Do not change any previous answers.)

Discussion

If you are a computer, you probably answered the questions as follows: joke, yoke, albumen, 50, not sufficient information, and yolk. If you are not a computer, you very likely answered some of the questions differently. If you have friends who are not computers, try these questions on them.

Human beings often look for quick and easy rules to solve problems. Sometimes those rules help. Sometimes they hinder. But we all do it. How quickly we are able to pick up a nonpertinent pattern and follow it! And how hard it is for us to give up those quick and easy rules even when we know better.

There is considerable evidence that the procedures described in this book for teaching children eliminate many of the differences we are accustomed to seeing between the mathematical learning of boys and girls, between the learning of "middle class" and "lower class" children, and between the learning of the groups we refer to as "minorities" and other children.

But beyond that, the way we treat individual children in the mathematics classroom has a great deal to do with how they see themselves mathematically. If girls are excused from working hard to really learn some difficult mathematics (because girls don't need that stuff) while boys are made to do the work, boys will expect to learn and will learn more mathematics. If Native Americans are assumed not to be good in mathematics, they will be not good in mathematics.

We are all more or less prejudiced. We have grown up in a sexist, racist society, and it has had its effect on us. If you think Wendy's mother must be the 2nd grade teacher because mathematicians don't become mothers (or vice versa), you are sexist. That's normal. But it's not good. It is especially not good if you are an educator who will influence the learning and the prejudices of the next generation. Even if you answered all the questions correctly, as an educator you must constantly take precautions not to treat children differently based on their sex, their ethnic background, their parents' occupations, or other nonpertinent information. If you didn't answer the questions correctly, you may have to be doubly careful.

Change is needed in the way all children learn mathematics. As a matter of equity, we should stop ignoring 90 percent of our population when we teach mathematics. Equally important for society, we cannot hope for the solution of the problems that will face us in the 21st century if we fail to educate all children to the limit of their capacity. In a world that is becoming steadily more quantitative, we must provide better mathematics education, for everyone, from kindergarten through graduate school.

All children learn and use mathematics better if it is derived from their reality, abstracted, practiced in enjoyable and effective ways, and applied to situations that are interesting and real to them. All children must learn to communicate more easily about and with mathematics. All children must learn to use technology efficiently to help them solve their problems.

Children must learn to think of technology as a tool or servant rather than something mystical of which they should be afraid or in awe. Children must learn to use their native intellect to solve real problems that involve mathematical thinking. They must learn in a way that will make them want to think mathematically, rather than in a way that will make them want to avoid mathematics at all costs.

Because of the ever increasing body of mathematical thought that is available and the ever wider uses of mathematics, perhaps the most important outcome of school mathematics should be that all people leave school with the ability and desire to continue learning more mathematics and to continue learning new ways to use mathematics.

References and Further Resources

Arbeiter, S. (1984). *Profiles, College-Bound Seniors, 1984.* New York: College Entrance Examination Board.

Committee on the Mathematical Education of Teachers (COMET) of the Mathematical Association of America (MAA). (1988). *Guidelines for the Continuing Mathematical Education of Teachers.* Notes #10. Washington, D.C.: MAA.

Economic Policy Institute. (1990). "Shortchanging Education: How U.S. Spending on Grades K–12 Lags behind Other Industrial Nations." Reported in *Education Week* (January 24, 1990). Pages 1,22. (Available for $2, prepaid, from the Economic Policy Institute, 1730 Rhode Island Ave., NW, Suite 812, Washington, DC 20036.)

Flanders, J.R. (1987). "How Much of the Content in Mathematics Textbooks Is New?" *Arithmetic Teacher* 35: 18–23.

Jacobs, H.R. (1970). *Mathematics, a Human Endeavor: A Textbook for Those Who Think They Don't Like the Subject.* San Francisco: W.H. Freeman.

Kenelly, J.W., ed. (1989). *The Use of Calculators in the Standardized Testing of Mathematics.* New York: The College Board and The Mathematical Association of America.

National Council of Teachers of Mathematics. (1984). *Professional Standards for Selection and Implementation of Instructional Materials.* Reston, Va.: NCTM.

National Council of Teachers of Mathematics. (1989a). *Curriculum and Evaluation Standards for School Mathematics.* Reston, Va.: NCTM.

National Council of Teachers of Mathematics. (1989b). "Professional Standards for Teaching Mathematics." (Working draft). Prepared by the Working Group of the Commission of Teaching Standards for School Mathematics of NCTM. Reston, Va.: NCTM.

Tyson-Bernstein, H. (1988). *A Conspiracy of Good Intentions: America's Textbook Fiasco.* Washington, D.C.: Council for Basic Education.

Will, G.F. (January 5, 1987). "Three Balls, Two Strikes." *Newsweek.*

Willoughby, S.S. (1984). "Mathematics Education 1984: Orwell or Well?" *Arithmetic Teacher* 32, 2: 54–59.